PROTECTIVE GLOVES for OCCUPATIONAL USE

EDITED BY

Gunh A. Mellström, P.D., Ph.D.
Department of Pharmaceutics
ACL National Corporation of Swedish Pharmacies
Stockholm, Sweden

Jan E. Wahlberg, M.D., Ph.D.
Department of Occupational Dermatology
Karolinska Hospital
Stockholm, Sweden

Howard I. Maibach, M.D.
Department of Dermatology
University of California
San Francisco, California

CRC Press
Boca Raton   Ann Arbor   London   Tokyo

Library of Congress Cataloging-in-Publication Data

Protective gloves for occupational use / edited by Gunh A. Mellström, Jan E. Wahlberg, Howard I. Maibach.
    p. cm. — (CRC series in dermatology)
    Includes bibliographical references and index.
    ISBN 0-8493-7359-X
    1. Contact dermatitis—Prevention. 2. Gloves. 3. Protective clothing. 4. Dermatotoxicology. 5. Industrial hygiene. 6. Occupational diseases. I. Mellström, Gunh A. II. Wahlberg, Jan E. III. Maibach, Howard I. IV. Series.
    [DNLM: 1. Protective Clothing—adverse effects. 2. Protective Clothing—standards. 3. Dermatitis, Occupational—etiology. 4. Hand Dermatoses—etiology. 5. Rubber—adverse effects. WR 140 P967 1994]
RL244.P76   1994
685'.43—dc20
DNLM/DLC
for Library of Congress   93-35532
                                                                                  CIP

    This book contains information obtained from authentic and highly regarded sources. Reprinted material is quoted with permission, and sources are indicated. A wide variety of references are listed. Reasonable efforts have been made to publish reliable data and information, but the author and the publisher cannot assume responsibility for the validity of all materials or for the consequences of their use.

    Neither this book nor any part may be reproduced or transmitted in any form or by any means, electronic or mechanical, including photocopying, microfilming, and recording, or by any information storage or retrieval system, without prior permission in writing from the publisher.

    All rights reserved. Authorization to photocopy items for internal or personal use, or the personal or internal use of specific clients, may be granted by CRC Press, Inc., provided that $.50 per page photocopied is paid directly to Copyright Clearance Center, 27 Congress Street, Salem, MA 01970 USA. The fee code for users of the Transactional Reporting Service is ISBN 0-8493-7359-X/94/$0.00+$.50. The fee is subject to change without notice. For organizations that have been granted a photocopy license by the CCC, a separate system of payment has been arranged.

    CRC Press, Inc.'s consent does not extend to copying for general distribution, for promotion, for creating new works, or for resale. Specific permission must be obtained in writing from CRC Press for such copying.

    Direct all inquiries to CRC Press, Inc., 2000 Corporate Blvd., N.W., Boca Raton, Florida 33431.

© 1994 by CRC Press, Inc.

No claim to original U.S. Government works
International Standard Book Number 0-8493-7359-X
Library of Congress Card Number 93-35532
Printed in the United States of America   3 4 5 6 7 8 9 0
Printed on acid-free paper

CRC Series in
DERMATOLOGY: CLINICAL AND BASIC SCIENCE
Edited by Dr. Howard I. Maibach

The CRC Dermatology Series combines scholarship, basic science, and clinical relevance. These comprehensive references focus on dermal absorption, dermabiology, dermatopharmacology, dermatotoxicology, and occupational and clinical dermatology.

The intellectual theme emphasizes in-depth, easy to comprehend surveys that blend advances in basic science and clinical research with practical aspects of clinical medicine.

Published Titles:

Hand Eczema
Torkil Menne and Howard I. Maibach

Health Risk Assessment: Dermal and Inhalation Exposure and Absorption of Toxicants
Rhoda G. M. Wang, James B. Knaak, and Howard I. Maibach

Pigmentation and Pigmentary Disorders
Norman Levine

Forthcoming Titles:

Bioengineering of the Skin: Water and the Stratum Corneum
Peter Elsner, Enzo Berardesca, and Howard I. Maibach

Bioengineering of the Skin: Cutaneous Blood Flow and Erythema
Enzo Berardesca, Peter Elsner, and Howard I. Maibach

Handbook of Contact Dermatitis
Christopher J. Dannaker, Daniel J. Hogan, and Howard I. Maibach

Human Papillomavirus Infections in Dermatovenereology
Gerd Gross and Geo von Krogh

Mouse Mutations with Skin and Hair Abnormalities: Animal Models and Biomedical Tools
John P. Sundburg

Skin Cancer: Mechanisms and Relevance
Hasan Mukhtar

The Contact Urticaria Syndrome
Arto Lahti and Howard I. Maibach

The Irritant Contact Dermatitis Syndrome
Pieter Van der Valk, Pieter Coenrads, and Howard I. Maibach

FOREWORD

More than 90% of occupational skin diseases are diagnosed as contact dermatitis. For this reason, the contents of this book focus on protective gloves for use against chemicals harmful to the skin and against microorganisms. Protection against mechanical and thermal risks, cold and fire, against ionizing radiation and radioactive contamination are not considered. Protective gloves are used in many situations: at work, at home, and in sparetime. In recent years there has been both an increased occupational use of protective gloves and increased interest in their protective capacity against harmful chemicals as well as blood-borne infections. This has resulted in intense work on the development of performance standard test methods. In order to make a correct selection it is necessary to have experimentally supported and evaluated information concerning the gloves' resistance to harmful chemicals and/or to microorganisms based on different kinds of test data and not only on the basis of cost and appearance. It is also important to be aware of the side effects which can arise when using protective gloves. The increased use of disposable rubber latex gloves has resulted in growing incidence of allergic reactions among health care personnel as well as patients treated by personnel wearing latex gloves.

The aim of this book is to explain briefly and simply, not only to dermatologists, but to everyone who has to use, select, or purchase protective gloves:

- Risk assessment of exposure to hazardous chemicals and infectious materials
- Rules and regulations concerning the use of protective gloves
- Standard quality control testing and testing of protective effect *in vitro,* in animals and clinical testing in patients
- Different kinds of side effects caused by glove materials
- Information sources of test data and their application in the selection process of protective gloves for different kinds of work situations

It is our hope that this book will be useful as a textbook as well as a reference book.

G. A. Mellström, Stockholm, Sweden
J. E. Wahlberg, Stockholm, Sweden
H. I. Maibach, San Francisco, California

THE EDITORS

Gunh A. Mellström, P.D., Ph.D., is Vice Manager of the Department of Pharmaceutics, Central Laboratory, National Corporation of Swedish Pharmacies, and Consultant Specialist at the Department of Occupational Dermatology, National Institute of Occupational Health in Stockholm, Sweden.

Dr. Mellström obtained her training at the University of Uppsala, Sweden, receiving the degree of P.D. (Pharmaceutical Doctor) in 1969 and a Ph.D. degree (in Medicine) at the Karolinska Institute, Stockholm in 1991.

She served as a lecturer in Pharmaceutics, at the Faculty of Pharmacy, University of Uppsala from 1968, and as Research Pharmacist at the Department of Occupational Dermatology, National Institute of Occupational Health from 1979 to 1989. It was in 1989 that she assumed her present positions.

Dr. Mellström is the Swedish representative member of the Board of Directors of NOKOBETEF (Nordic Coordination Group on Protective Clothing as a Technical Preventive Measure) and a member of the Swedish Academy of Pharmaceutical Sciences.

Dr. Mellström is the author of more than 40 papers and co-author of 2 books. Her current research interest relates to the protective effects of gloves against antineoplastic drugs.

Jan E. Wahlberg, M.D., Ph.D., is a Professor in the Department of Occupational Dermatology at the National Institute of Occupational Health and Karolinska Hospital, Stockholm, Sweden.

He obtained his M.D. from the Karolinska Institute, Stockholm in 1959 and his Ph.D. in 1966 at the same university for investigations on percutaneous absorption of metal compounds.

Dr. Wahlberg is a member of the Swedish Dermatological Society, Chairman of the International Contact Dermatitis Research Group (ICDRG), and a member of the European Environmental Contact Dermatitis Research Group.

His research includes occupational dermatology, contact dermatitis, and dermatotoxicology. He has published more than 200 research and review papers.

Howard I. Maibach, M.D., is Professor of Dermatology, School of Medicine, University of California, San Francisco. Dr. Maibach graduated from Tulane University, New Orleans, Louisiana (A.B. and M.D.) and received his research and clinical training at the University of Pennsylvania, Philadelphia. He received an honorary doctorate from the University of Paris Sud in 1988.

Dr. Maibach is a member of the International Contact Dermatitis Research Group, the North American Contact Dermatitis Group, and the European Environmental Contact Dermatitis Group. He has published more than 1100 papers and 40 volumes.

CONTRIBUTORS

Michael H. Beck, FRCP
Skin Hospital
Salford, England

Stephen P. Berardinelli, Sr., Ph.D.
Centers for Disease Control
National Institute of Occupational
 Safety and Health
Morgantown, West Virginia

Anders S. Boman, Ph.D.
Department of Occupational
 Dermatology
National Institute of Occupational
 Health
Solna, Sweden

Lars G. Burman, M.D., Ph.D.
Swedish Institute for Infectious
 Disease Control
Stockholm, Sweden

Harry F. Bushar, Ph.D.
FDA Center for Devices
 and Radiological Health
Rockville, Maryland

Ronald F. Carey, Ph.D.
FDA Center for Devices
 and Radiological Health
Rockville, Maryland

Birgitta Carlsson
Engineer
National Board of Occupational
 Safety and Health
Solna, Sweden

W. Howard Cyr, Ph.D.
FDA Center for Devices
 and Radiological Health
Rockville, Maryland

J. G. Dillon, Ph.D.
Office of Science and Technology
FDA Center for Devices
 and Radiological Health
Rockville, Maryland

**An E. Dooms-Goossens,
 R. Pharm., Ph.D.**
Department of Dermatology
University Hospital St. Rafaël
Leuven, Belgium

Tuula Estlander, M.D., Ph.D.
Department of Dermatology
Institute of Occupational Health
Helsinki, Finland

Birgitta Fryklund, R.N.
Swedish Institute for Infectious
 Disease Control
Stockholm, Sweden

Curtis P. Hamann, M.D.
SmartPractice
Phoenix, Arizona

Angelika Heese, M.D.
Department of Dermatology
University Hospital
Erlangen, Germany

Norman W. Henry, III, M.S.
DuPont Company
Newark, Delaware

Bruce A. Herman, M.S.
Office of Science and Technology
FDA Center for Devices
  and Radiological Health
Rockville, Maryland

Otto Paul Hornstein, M.D.
Department of Dermatology
University Hospital
Erlangen, Germany

Riitta Jolanki, D. Tech.
Department of Dermatology
Institute of Occupational Health
Helsinki, Finland

Lasse Kanerva, M.D., Ph.D.
Department of Dermatology
Institute of Occupational Health
Helsinki, Finland

Shelley A. Kick, Ph.D.
SmartPractice
Phoenix, Arizona

Hans Uwe Koch, Ph.D.
Department of Dermatology
University Hospital
Erlangen, Germany

Helen J. Rosen Kotilainen, M.A., M. Tech. (ASCP), CIC
Infection Control Department
University of Massachusetts
  Medical Center
Worcester, Massachusetts

Paul Leinster, M.B.A., Ph.D.
Thomson-MTS Ltd.
Milton Keynes, England

Carola Lidén, M.D., Ph.D.
Department of Occupational
  Dermatology
Karolinska Hospital and
National Institute of
  Occupational Health
Stockholm, Sweden

Christopher R. Lovell, M.D., MRCP
Department of Dermatology
Royal United Hospital
Bath, England

C. David Lytle, Ph.D.
FDA Center for Devices
  and Radiological Health
Rockville, Maryland

Howard I. Maibach, M.D.
Department of Dermatology
University of California
School of Medicine
San Francisco, California

S. Z. Mansdorf, Ph.D., CIH, CSP
Mansdorf & Associates, Inc.
Stow, Ohio

Gunh A. Mellström, P.D., Ph.D.
Department of Pharmaceutics
ACL National Corporation of
  Swedish Pharmacies and
Department of Occupational
  Dermatology
National Institute of Occupational
  Health
Stockholm, Sweden

Klaus-Peter Peters, M.D.
Department of Dermatology
University Hospital
Erlangen, Germany

Leroy W. Schroeder, Ph.D.
Office of Science and Technology
FDA Center of Devices
  and Radiological Health
Rockville, Maryland

David G. Shombert, M.S.
Office of Science and Technology
FDA Center of Devices
  and Radiological Health
Rockville, Maryland

James S. Taylor, M.D.
Department of Dermatology
Cleveland Clinic Foundation
Cleveland, Ohio

Kristiina Turjanmaa, M.D., Ph.D.
Department of Dermatology
Tampere University Hospital
Tampere, Finland

Jan E. Wahlberg, M.D., Ph.D.
Department of Occupational
  Dermatology
Karolinska Hospital and
  National Institute of
  Occupational Health
Stockholm, Sweden

Karin Wrangsjö, M.D., Ph.D.
Department of Occupational
  Dermatology
Karolinska Hospital
Stockholm, Sweden

TABLE OF CONTENTS

Introduction

Introduction
Terminology and Abbreviations 3

Chapter 1.
Prevention of Contact Dermatitis 7
Jan E. Wahlberg and Howard I. Maibach

Chapter 2.
Industrial Hygiene Assessments for the Use of Protective Gloves 11
S. Z. Mansdorf

Chapter 3.
Gloves: Types, Materials, and Manufacturing 21
Gunh A. Mellström and Anders S. Boman

Rules and Regulations

Chapter 4.
European Standards on Protective Gloves 39
Gunh A. Mellström and Birgitta Carlsson

Chapter 5.
Protective Gloves for Occupational Use — U.S. Rules, Regulations,
and Standards 45
Norman W. Henry, III

Testing of Protective Effect

Chapter 6.
Testing of Protective Effect Against Liquid Chemicals 53
Gunh A. Mellström, Birgitta Carlsson, and Anders S. Boman

Chapter 7.
Chemical Resistance Field Test Methods 79
Stephen P. Berardinelli, Sr.

Chapter 8.
Percutaneous Absorption Studies in Animals 91
Anders S. Boman and Gunh A. Mellström

Chapter 9.
Standard Quality Control Testing and Virus Penetration 109
**C. David Lytle, W. Howard Cyr, Ronald F. Carey,
David G. Shombert, Bruce A. Herman, James G. Dillon,
Leroy W. Schroeder, Harry F. Bushar, and Helen J. Rosen Kotilainen**

Clinical Tests With Protective Gloves

Chapter 10.
Diagnosis-Driven Management of Natural Rubber Latex
Glove Sensitivity 131
Curtis P. Hamann and Shelley A. Kick

Chapter 11.
Clinical Testing for Rubber Glove Allergy 157
Michael H. Beck and Christopher R. Lovell

Chapter 12.
Clinical Testing of Occupation-Related Glove Sensitivity 171
An E. Dooms-Goossens

Chapter 13.
Allergologic Evaluation and Data on 173 Glove-Allergic Patients 185
**Angelika Heese, Klaus-Peter Peters, Hans Uwe Koch,
and Otto Paul Hornstein**

Chapter 14.
Protective Effect of Gloves Illustrated by Patch
Testing — Practical Aspects 207
Carola Lidén and Karin Wrangsjö

Side Effects When Using Protective Gloves

Chapter 15.
Irritation and Contact Dermatitis from Protective
Gloves — An Overview 215
Jan E. Wahlberg

Chapter 16.
Allergic Contact Dermatitis from Rubber and Plastic Gloves 221
Tuula Estlander, Riitta Jolanki, and Lasse Kanerva

Chapter 17.
Contact Urticaria from Latex Gloves 241
Kristiina Turjanmaa

Chapter 18.
Other Reactions from Gloves 255
James S. Taylor

Application of Test Data

Chapter 19.
The Selection and Use of Gloves Against Chemicals 269
Paul Leinster

Chapter 20.
The Selection and Use of Gloves by Health Care Professionals 283
Lars G. Burman and Birgitta Fryklund

Chapter 21.
Information Sources for Glove Test Data 293
Gunh A. Mellström

Index 305

PART I
INTRODUCTION

INTRODUCTION
Terminology and Abbreviations

I. TERMINOLOGY

A protective glove is an item of personal equipment which protects the hand or any part of the hand from hazards. It may also cover part of the forearm. Some of the terms used in technical permeation testing and in the dermatological field may need some clarification.

Penetration of a chemical is a process which can be defined as the flow of a chemical through closures, porous materials, seams, and pinholes or other imperfections in a protective clothing material and on a nonmolecular level. It should be pointed out that in experimental dermatology the term penetration is used with a somewhat different meaning.

Percutaneous absorption is defined as the penetration of substances from outside into the skin and thereafter through the skin and into the blood and lymph vessels. The three routes of penetration are (1) via the sweat ducts, (2) via the hair follicles and the associated sebaceous glands (also called transappendageal transport) and can be considered equivalent with the definition of penetration used in connection to material testing; and (3) across the continuous stratum corneum (also called transepidermal transport) and corresponds to the definition for permeation described below.

Permeation in technical testing means the process by which a chemical migrates through the protective glove material on a molecular level. It involves (1) sorption of molecules of the chemical into the contacted surface of the material, (2) diffusion of the sorbed molecules within the material, and (3) desorption of the molecules from the opposite surface of the material.

Chemical resistance of a protective material can also be tested *in vivo* using experimental animals and in clinical studies on patients as a part of making a diagnosis. For those not familiar with these test methods a short description is given below.

Percutaneous absorption studies in animals can be performed in the following way. A skin area on the animal (e.g., guinea pig) is exposed to the test chemical in a closed depot on top of a glove membrane and as a control without any protective membrane. The capacity of the membrane to reduce the percutaneous absorption of the test chemical in comparison with the unprotected site is determined by measuring the concentration in regularly taken blood samples. This kind of investigation, however, is not used as a routine method for glove testing.

Clinical studies of the protective effect of gloves can be studied by patch testing techniques. A piece of the glove is placed between the skin and the test substances, usually on the back of the patients. The results are read as the difference in reactivity between protected and unprotected skin.

II. ABBREVIATIONS

Some commonly used abbreviations are listed below. Most abbreviations for chemical names used in the text are explained when first mentioned there and are not given in this list.

AIDS — Acquired immune deficiency syndrome
AQL — Acceptable quality level
ASTM — American Society for Testing and Materials
BOHS — British Occupational Hygiene Society
BT, BTT — Breakthrough time
CD — Color developing agent
CDC — Centers for Disease Control
CEN — Comité Européen de Normalisation
CPC — Chemical protective clothing
CU — Contact urticaria
D — Dalton, kD — kilodalton
EECDRG — European Environmental and Contact Dermatitis Research Group
FDA — Food and Drug Administration
HBV — Hepatitis B virus
HCW — Health care workers
HIV — Human immunodeficiency virus
HSV — Herpes simplex virus
ICD — Immediate contact dermatitis
ICDRG — International Contact Dermatitis Research Group
ISO — International Standard Organization
MW — Molecular weight
NACDG — North America Contact Dermatitis Group
NICU — Nonimmunological contact urticaria
NRL — Natural rubber latex
PE — Polyethylene, polyethen
PPE — Personal protective equipment
PR, PER — Permeation rate
PVA — Polyvinylalcohol
PVAc — Polyvinylacetate
PVC — Polyvinylchloride
RAR — Relative absorption rate
RAST — Radioallergosorbent test

RPR — Relative permeation rate
RSV — Respiratory syncytical virus
SEM — Scanning electron microscopy
SPT — Skin prick test
TECP — Totally encapsulated chemical suits

1

Prevention of Contact Dermatitis

Jan E. Wahlberg and Howard I. Maibach

TABLE OF CONTENTS

I. Introduction 7
II. Protective Gloves 8
References 9

I. INTRODUCTION

A distinction is usually made between *primary prevention,* i.e., inhibition of the induction and onset of contact dermatitis, and *secondary prevention,* i.e., inhibition of relapses. The value of disease prevention is evident to man, society, and the medical community. For human, social, and economic reasons, it would be of great benefit if people exposed to harmful chemicals and products could be protected from developing contact dermatitis.

The multiple prophylactic means available are summarized in Table 1; they are grouped under subheadings: Chemical, Individual, Avoidance of Contact, Skin Care Program, and finally, Miscellaneous Means. It can be used as a checklist. These aspects on prevention have been dealt with in recent reviews.[1-6]

When facing the current or imminent skin problems of a single patient, all of these prophylactic means should be considered. However, the responsibility for primary prevention rests mainly with manufacturers and producers of chemicals and products, government agencies, consumer organizations, industrial physicians and nurses, and safety engineers.

Speaking of secondary prevention, a greater responsibility is placed on physicians treating the cases (dermatologists, industrial physicians, and others) and on nurses and safety engineers.

TABLE 1 Prophylaxis of Contact Dermatitis

Chemicals
    Identification of the allergen
    Occurrence, concentration of the allergen in the environment
    Allergen removal or replacement
    Modification or inactivation of the allergen
    Predictive testing: skin irritating potential
    Predictive testing: sensitizing potential
Individual identified at pre-employment examination and periodic health screening
    Those with increased susceptibility or predisposition, i.e., atopics
    Patients with a history of contact dermatitis
    Patients with stasis eczema and/or venous leg ulcers
Avoidance of direct contact with products and materials
    Protective gloves
    Aprons, sleeves, boots, glasses, masks, etc.
    Protective (barrier) creams
    Dishwasher, washing machine, long-handled brushes
    Automation, closed systems
    Efficient ventilation
Skin care program
    Soaps, detergents, cleansing agents with low irritancy potential
    Hot water, shower, sauna
    Soft towels
    Emollient and moisturizing creams
Miscellaneous
    Legislation
    Labeling of products and chemicals; safety sheets
    Information to patients, consumers, workers, supervisors
    Training of workers in special industrial processes
    Good housekeeping
    Research on prevention and dissemination of results obtained

II. PROTECTIVE GLOVES

The use of protective gloves is then one of several possibilities to avoid developing a contact dermatitis or a relapse. All the items listed in Table 1 should be considered, and, according to our experience, the best results are achieved when several of the recommended prophylactic means are combined in a wise and fruitful way. To rely on just one of these recommendations — sometimes to reduce costs — is definitely less effective and approaches malpractice! However, it is up to the persons involved in preventive dermatology to demonstrate that the suggested methods and measures are efficacious and cost effective.

Current protective gloves are not perfect. As documented in the following chapters, some are permeable to various chemicals and do not provide the promised protection. Side effects, such as irritancy, latex and contact allergy, are common and are sometimes reasons for discontinuance of their use by patients and exposed workers.

Gloves that will give more efficient protection and less side effects are then highly desirable. We are optimistic that devoted people from industry, universities, and research institutes will meet this challenge and, in the near future, present us with the desired products.

REFERENCES

1. Wahlberg, J. E., Prophylaxis of contact dermatitis, *Semin. Dermatol.,* 5 (3), 255, 1986.
2. Fisher, A. A. and Adams, R. M., Occupational dermatitis, in *Contact Dermatitis,* 3rd ed., Fisher, A. A., Ed., Lea & Febiger, Philadelphia, 1986, chap. 28.
3. Adams, R. M., Prevention, rehabilitation, treatment, in *Occupational Skin Disease,* Adams, R. M., Ed., W. B. Saunders Company, Philadelphia, 1990, chap. 16.
4. Rycroft, R. J. G., Occupational dermatoses, in *Textbook of Dermatology,* Vol. 1, 5th ed., Champion, R. H., Burton, J. L., and Ebling, F. J. G., Eds., Blackwell Scientific Publications, Oxford, 1992, chap. 18.
5. Rycroft, R. J. G., Occupational contact dermatitis, in *Textbook of Contact Dermatitis,* Rycroft, R. J. G., Menné, T., Frosch, P. J., and Benezra, C., Eds., Springer-Verlag, Berlin, 1992, 341.
6. Lachapelle, J.-M., Principles of prevention and protection in contact dermatitis (with special reference to occupational dermatology), in *Textbook of Contact Dermatitis,* Rycroft, R. J. G., Menné, T., Frosch, P. J., and Benezra, C., Eds., Springer-Verlag, Berlin, 1992, 695.

2

Industrial Hygiene Assessments for the Use of Protective Gloves

S. Z. Mansdorf

TABLE OF CONTENTS

I. Introduction 11
II. The Industrial Hygiene Process 12
III. Assessing Risk 12
    A. Frequency and Severity 13
    B. Exposure and Dose 13
IV. TLVs and Other Published Exposure Recommendations 14
V. Assessment of Risk 15
VI. Levels of Risk 17
VII. Consideration of Other Control Measures 17
VIII. Risk vs. Benefit 17
IX. Worker Training 18
References 19

I. INTRODUCTION

Prevention of work-related diseases and injuries affecting the skin requires an assessment of the hazards and the risks associated with performing the work prior to the assignment of protective clothing or equipment. This assessment should consider the nature and extent of the hazards involved, the likelihood of exposure to hazardous materials or agents, the potential severity of adverse effects of exposure to the worker, options other than personal protective equipment use to protect the worker, and the overall risk to the user performing the task or work while using protective equipment. The industrial hygiene process incorporates these assessments.

II. THE INDUSTRIAL HYGIENE PROCESS

The focus of the art and science of industrial hygiene is the anticipation, recognition, evaluation, and control of chemical and physical agents and other stressors in the workplace and community.[1] Anticipation of health and safety hazards based solely on a review of plans for a process, on proposed changes in materials, or on job descriptions alone requires the most sophisticated level of industrial hygiene knowledge and skill. As might be suspected, this skill relies heavily on a mix of experience and knowledge to anticipate potential problem areas before they actually occur. The next aspect of the industrial hygiene process — recognition of a problem — is a somewhat less difficult task, although one that still relies a great deal on both experience and knowledge. Many times this phase of industrial hygiene work is focused on employee complaints, reports of injury or illness, and other concrete indicators of potential adverse health effects. Evaluation of the problem follows the recognition phase. Industrial hygiene evaluations can be either qualitative or quantitative or both. For materials which present a skin hazard, the evaluation will usually be completed on a qualitative basis, since there are no standard measures for quantifying dermal hazards in the workplace.[2-3] The major emphasis of the evaluation phase is the assessment of risk from exposure, followed by the subsequent consideration of controls available to limit the risk to a point that is considered acceptable by both the employer and the worker. One control method common to many applications is the use of personal protective equipment, especially gloves for the protection of the skin and the hands. For most jobs, it is the hands that are most likely to suffer the consequences of contact with a hazardous chemical or physical agent. Further, organizations such as the National Institute for Occupational Safety and Health have recognized skin diseases and disorders as a major occupational health problem.[4] Therefore, gloves have a significant role in protecting workers from chemical and physical hazards.

III. ASSESSING RISK

Risk can be defined for purposes of this discussion as the likelihood of an undesirable effect. In our application, this is most typically an adverse health effect. Risk is usually based on the severity of effects of contact and either the likelihood, duration, or frequency of contact. While there are other approaches and factors that can also play a part in the determination of relative risk, the major factors are severity and frequency. An example of extreme severity from the petrochemical industry can be used to illustrate this point. The greatest hazard by far of working in a refinery is the danger of fire and explosion. Major fires are an extremely rare event (very low frequency), but the severity is so great that most companies are willing to go to great lengths and expense to protect against it. This has included the mandated use of fire resistant clothing by all workers as well as other measures. The importance of length of contact (duration) or frequency can be shown by the example of thermal burns to the skin. The key determinate for severity of burn is the length of time of contact with a hot surface once the temperature is above 45°C (113°F).[5-6] The longer the contact period, the greater the likelihood of a burn or scalding. In this example, the risk increases with the contact time.

A. Frequency and Severity

Our final example of the role of frequency and severity in determining risk will be illustrated by the routine task of putting gasoline into a car. Gasoline is universally recognized as a hazardous compound. It is extremely flammable and clearly toxic. It contains a wide range of organic constituents, some of which probably exhibit skin permeability, toxicity, and carcinogenicity (e.g., benzene). Most would not argue these points, yet few of us use gloves when putting gasoline into our cars. Even the gas station attendants rarely use gloves. There are probably several reasons why gloves are not worn, including the matter of convenience. Nevertheless, a key aspect is that we consider the secondary contact (i.e., contact from the dispensing nozzle not the actual fluid) and frequency of exposure to represent a trivial risk. This personal risk assessment is also greatly aided by the fact that gasoline is a familiar product with which almost everyone has had some experience. Nevertheless, we do not usually apply this logic at work. If the components of the gasoline mixture were separated and labeled, most of us would want to wear gloves even though the exposure potential might be trivial. As Grandjean states in the conclusion to *Skin Penetration: Hazardous Chemicals at Work*, " ... skin contact with a known skin hazard is not always a danger."[7]

B. Exposure and Dose

The calculation of exposure dose is relatively straightforward and simply a function of surface area exposed, concentration of the chemical, and exposure duration. However, the uptake rate through the skin and subsequent pharmacokinetics are much more difficult to define.[8] Even if we compute a quantity that is on the skin, it may have little current value. Acceptable dose values for skin exposures have not been established except for a few isolated chemicals, and these data are usually a result of pharmacological studies for drugs (e.g., transdermal drug delivery, such as the controlled release skin patches containing nitroglycerin for the control of angina).[9]

In the case of airborne exposures, most of the data presently used for determining safe airborne exposure concentrations are based on years of epidemiological data from worker populations. Studies of the health effects of exposure where the major route of entry is dermal are lacking.[3] Second, it would be difficult to establish an acceptable level of exposure where no effective means is available to measure it. This is not the case with airborne exposures. Because there are no "acceptable" exposures or doses established for the dermal route, most workers and health and safety professionals have used a no or zero exposure goal. This results in the selection and use of gloves and other clothing for protection against skin exposures that would exhibit breakthrough resistance for at least as long as the job or task lasts.[10] The problem with this approach is that it does not consider cost/benefit factors, assumes no threshold (linear approach) of adverse effects, and is not feasible in many situations, since impermeable protective clothing does not presently exist.

IV. TLVs AND OTHER PUBLISHED EXPOSURE RECOMMENDATIONS

Threshold Limit Values (TLVs) for Chemical Substances and Physical Agents, published by the American Conference of Governmental Industrial Hygienists (ACGIH), and the Occupational Safety and Health Administration's Permissible Exposure Limits (PELs), published under 29 CFR 1910.1000 Table Z-1-A, provide either recommended or legally mandated exposure limits for airborne chemical substances. In the evaluation of inhalation hazards, the industrial hygienist has the benefit of either the TLVs or PELs to use as a benchmark of relative risk. They can compare the values found through air sampling on a quantitative basis to what is considered either "safe" or acceptable for the average worker. Each organization also provides a "skin" notation for those chemicals believed to present a significant risk of skin absorption and consequent toxicity. Examples of these chemicals include carbon disulfide, carbon tetrachloride, and the common solvent, methyl alcohol. Nevertheless, the American Conference of Governmental Industrial Hygienists states in their own Documentation of Biological Exposure Indices that " ... quantitative data on dermal absorption of industrial chemicals are scattered, and dermal notations are inconsistent ... studies indicate that a contribution of dermal exposure to the total dose is more common than is indicated by a skin notation."[11] These tables also do not indicate those chemicals capable of causing chemical burns or corrosion of the skin. This is a very important risk consideration. A special notation of "caution" is used in the *Quick Selection Guide to Chemical Protective Clothing,* 2nd ed.; however, it is not entirely unique to corrosion but rather a broader measure of risk of adverse effects on the skin to include corrosion.[12] Other indicators of risk include the classification scheme of the European Community Gazette, where they publish a "risk code" in their "Guide to Classification and Marking of Hazardous Substances." This listing includes the categories of highly corrosive, corrosive, and irritants.[12] They also note those chemicals which are known sensitizers.

Because there are no TLVs or PELs for the skin but simply forms of warning, the risk of exposure has to be determined for each substance based on the way that it is used and the inherent toxicological risk of the substance itself. Biological monitoring can be used to determine the actual dose as part of a medical surveillance program; however, most substances do not have accurate biological markers, and in all cases the results are measures of the dose already received (and the damage already done). Given these circumstances, the risk should be assessed based upon the exposure potential of the work assignment.

V. ASSESSMENT OF RISK

The first step in the risk assessment process is the characterization of the exposure. This is accomplished by an evaluation of the process or work task, the exposure conditions, and the physical, chemical, biological, or radiological hazards present.[13]

The work being evaluated should be broken down into individual elements or the steps required to complete the task. This process is essentially the same as that used in Job Safety Analysis or the more traditional Timestudy Analysis. Each distinct aspect of the process or task should be broken down into finite steps. A relatively simple large industrial battery charging operation will be used as an example:

1. Put on protective clothing consisting of a Neoprene® apron, Neoprene® gloves, and a faceshield (estimated average duration of 2 min).
2. Use an overhead hoist to transport a battery from the receiving rack to the charging area (estimated average duration of 5 min).
3. Remove a battery cap and obtain a sample to check the specific gravity of the battery acid (estimated average duration of 2 min).
4. Add battery acid (sulfuric) as required using a bulk dispenser hose with squeeze spigot (2 to 10 min estimated average duration).
5. Clean tools by rinsing in water (estimated average duration of 2 min).
6. Clean battery top using baking soda and water and a brush (estimated average duration of 4 min).
7. Place battery on charge by connecting terminals (estimated average duration of 2 min).
8. Remove protective clothing or repeat for another battery.

Following the description of distinct tasks within the overall job, a listing of chemical, physical, biological, radiological, or other hazards needs to be developed for each step in the task.

For chemical exposures, the temperature, concentration of the compound, and the type of vehicle or carrier are all important factors. Second, the type of contact and duration or frequency are also of importance. There can be a wide variation in the potential for contact with a hazardous chemical. Examples of the types of contact are

- Continuous direct contact with solids
- Intermittent direct contact with solids
- The potential for direct contact with solids
- Continuous direct liquid contact through immersion
- Intermittent direct liquid contact through immersion

- Intermittent direct liquid contact through splashing, spraying, or misting
- The potential for direct liquid contact
- Continuous vapor contact
- Intermittent vapor contact
- The potential for vapor contact

The type of contact or potential for contact will have a major impact on the overall risk the work task presents.

Physical hazards and other special needs for gloves or protective clothing to provide either resistance to or protection from physical hazards are also important to our characterization of the risk of the job. This information can also be used to help determine the risk of the job while using protective equipment. Examples of the types of physical hazards to consider are

- Tears, cuts, and punctures to the skin or glove
- Abrasion of the skin or glove
- Thermal resistance and protection
- Flammability of the protective equipment
- Protection from radiation

Using our example again, the following exposure assessment analysis is developed:

1. Put on protective clothing consisting of a Neoprene® apron, Neoprene® gloves, and a face shield (requires inspection of protective equipment; face shield may limit visibility).
2. Use an overhead hoist to transport a battery from the receiving rack to the charging area (potential for exposure to battery acid by splash contact or by physical contact with the battery housing; requires the protective gloves to have a moderate level of abrasion resistance).
3. Remove a battery cap and obtain a sample to check the specific gravity (potential for direct liquid contact with battery acid; this requires gloves which do not have a slippery surface when wet and permit enough dexterity to remove the battery cap and to handle the instrument).
4. Add battery acid (sulfuric) as required using a bulk dispenser hose with squeeze spigot (potential for direct liquid contact with the glove by acid and potential for splash).
5. Clean tools by rinsing in water (requires gloves which will permit water washing).
6. Clean battery top using baking soda and water and a brush (requires gloves which have good puncture and tear resistance).
7. Place battery on charge by connecting terminals (requires gloves to have a moderate level of cut resistance; electrical resistance is also of benefit).
8. Remove protective clothing or repeat for another battery (requires decontamination to remove any sulfuric acid that may have contaminated the gloves or other items; also requires gloves to be capable of being water washed and air dried and stored without any degradation).

VI. LEVELS OF RISK

We can categorize risk in many different ways using subjective or objective criteria. One approach is to group the hazards from chemicals and physical agents into levels of risk. As an example, a numeric scoring scheme ranging between one to five can be used where the number one would represent the greatest risk and the number five the lowest risk. Contact with anhydrous hydrofluoric acid might represent a risk level of one, while contact with tap water might represent a risk level of five. Using this scheme, a level five risk might permit use of a glove which could or does allow permeation, while a risk level one application would require the use of a glove that offers complete permeation resistance for the length of the operation. In our example of the battery charging operation, this might represent a moderate (e.g., level three) level of risk since the hazard is acute and the acid is somewhat diluted. This means that a failure of the protective device (i.e., glove) would not likely result in a serious permanent injury. Hence, we would probably consider the use of protective gloves for this work task as being acceptable. By using this "acceptable risk concept", we can assign gloves which might exhibit permeation at low levels for certain tasks. An example is the application of isopropyl for cleaning, since the major effect of failure would be a mild dermatitis from defatting of the skin.

VII. CONSIDERATION OF OTHER CONTROL MEASURES

Before assigning workers protective gloves or other personal protective equipment in high risk jobs, other control measures should be considered. These include the following:

- Changes in work practices, such as the use of tools
- Engineering controls or other process changes
- Robotics and other forms of automation

A key factor in the industrial hygiene and risk assessment process is the evaluation of actual work practices. There may be effective alternatives to the actual hand manipulation of an item or work task which are much less expensive and also less difficult to implement than engineering controls. Gloves and other forms of personal protective equipment should be assigned after we have evaluated the risk to the user in case of a failure. For those jobs where the risk of serious injury remains after assigning the protective clothing, we will need to consider other options.

VIII. RISK VS. BENEFIT

In some applications, the risk from wearing the gloves may outweigh the risk of not wearing gloves. One common situation where workers do not want to wear gloves is in the processing of specialty paints and inks on roller mills. The mixing of pigments for certain viscous paints and inks is done using inward rolling mills. The paints and inks have a solvent base and the pigments can stain the skin; however, mill safety precludes the use of polymer gloves which could get caught and pull a worker's hand into the mill. When gloves are worn, they are usually made of cloth and are very loose fitting so that they will allow for the worker to pull their hand out quickly and easily in case of an emergency. While there are other approaches to resolve this apparent problem, it does illustrate a common situation where the glove use is more dangerous than the limited exposure to the solvents and pigments.

A second example where glove applications require a risk/benefit evaluation is the present dilemma faced by the surgeon. There are gloves that are essentially cut proof and combinations (such as using multiple layers of gloves) available that would result in a glove system that is both cut proof and puncture resistant, thus providing complete protection against blood-borne pathogens. However, they would not permit the surgeon enough dexterity for even the most simple task. In this case, there is no product currently available that will accomplish the function desired by the surgeon, so there must be a trade-off between personal risk to the doctor and risk to the patient from reduced dexterity.

IX. WORKER TRAINING

Worker information and training are a very important part of the risk reduction process. Users should be provided with the following information:

- Nature and extent of the hazards
- Signs and symptoms of overexposure
- Use and limitations of the protective equipment assigned
- Decontamination procedures, if required
- Inspection, maintenance, and storage procedures
- First aid and emergency procedures

The industrial hygiene evaluation of the job should provide the information necessary to inform the worker of the nature and extent of the hazards and the level of risk the worker will face. This should include the specific nature of the hazard (e.g., corrosive, flammable, toxic, etc.) and the extent of the hazard. Other important information includes the signs and symptoms of overexposure so that the worker knows if the protective equipment has failed. The proper use and limitations of the protective equipment are important to provide the worker with the limits to the range of protection provided. How or when to decontaminate protective equipment is important in those situations where the equipment is likely to become contaminated and before reuse of the equipment. Inspection of the gloves prior to use by the employer will greatly decrease the potential for accidents. Prior to each use, the worker should also inspect the gloves for imperfections, discolorations, etc. If the gloves are clean, they should be inflated by blowing into them or by quickly folding them at the opening (leaving some air in the glove) and immersing them in water. Bubbles are an indication of pinholes or other discontinuities. Previously contaminated gloves should not be used before they are decontaminated. Finally, it is important to ensure that the worker knows what to do in the event of an emergency.

Knowledge of the hazard not only reduces the risk but also provides the health and safety professional with a worker capable of providing feedback on the effectiveness of the protective equipment.

REFERENCES

1. Plog, B. A., *Fundamentals of Industrial Hygiene,* National Safety Council, Chicago, 1988, 1-28.
2. Perkins, J. L., Chemical protective clothing. I. Selection and use, *Appl. Indust. Hyg.,* 2, 222, 1987.
3. Jacobs, R. R., Overview of skin structure, function, and toxicity, in *Chemical Protective Clothing,* Johnson, J. S. and Anderson, K. J., Eds., American Industrial Hygiene Association, Akron, OH, 1990, chap. 2.
4. NIOSH, Prevention of leading work related diseases and injuries, *MMWR,* 35, 1986.
5. Hughes, D., Contact injuries, in *The Thermal Environment,* BOHS Technical Guide No. 8, Science Reviews Ltd., H and H Scientific Consultants Ltd., Leeds, 1990, chap. 7.
6. Ripple, G. R., Torrington, K. G., Phillips, Y. Y., Predictive criteria for burns from brief thermal exposures, *JOM,* 32, 215, 1990.
7. Grandjean, P., Conclusions, in *Skin Penetration: Hazardous Chemicals at Work,* Taylor and Francis, New York, 1990, 179-180.
8. Mansdorf, S. Z., Industrial hygiene and safety needs for protective clothing, in *Proc. 4th Scand. Symp. Protective Clothing Against Chemicals and Other Health Hazards,* Makinen, H., Ed., NOKOBETEF, Helsinki, 1992, 1-10.
9. Mansdorf, S. Z., Risk assessment of chemical exposure hazards in the use of chemical protective clothing — an overview, in *Performance of Protective Clothing,* ASTM STP 900, Barker, R. L. and Coletta, G. C., Eds., American Society for Testing and Materials, Philadelphia, 1986, 207-213.
10. Roder, M. M., *A Guide for Evaluating the Performance of Chemical Protective Clothing,* NIOSH Publication 90-109, National Institute for Occupational Safety and Health, Cincinnati, 1990.
11. ACGIH, Dermal absorption, in *Documentation of the Threshold Limit Values and Biological Exposure Indices,* Vol. III, 6th ed., American Conference of Governmental Industrial Hygienists, Inc., Cincinnati, 1991, BEI-20.
12. Forsberg, K. and Mansdorf, S. Z., *Quick Selection Guide to Chemical Protective Clothing,* 2nd ed., Van Nostrand Reinhold, New York, 1993.
13. Johnson, J. S., Personal protective equipment selection criteria and field use, in *Chemical Protective Clothing,* Johnson, J. S. and Anderson, K. J., Eds., American Industrial Hygiene Association, Akron, OH, 1990, chap. 7.

3

Gloves: Types, Materials, and Manufacturing

Gunh A. Mellström and Anders S. Boman

TABLE OF CONTENTS

    I. Types 21
   II. Materials 22
       A. Rubber Materials (Natural and Synthetic) 23
       B. Plastic Polymer Materials 24
  III. Manufacturing 25
       A. Gloves Made by Dipping 25
           1. Compounding of Rubber Latex 25
           2. Vulcanizing Systems 26
           3. Dipping Processes 28
           4. Leaching, Chlorination, and Lubrication 30
           5. Drying, Curing, and Stripping 31
           6. Polymer Blends and Polymer Composites 31
       B. Supported Gloves 32
       C. Gloves Made by Sewing and Knitting 32
       D. Gloves Made by Punching and Welding (Foil Gloves) 33
       E. Summary 33
References 34

I. TYPES

Personal protective equipment — in this context, gloves — is an essential factor in preventing or reducing direct skin contact to harmful agents. For practical purposes, protective gloves can be classified into different types based on use and thickness.[1]

   Type I: Disposable gloves (thickness: 0.007 to 0.25 mm)
   Type II: Household gloves (thickness: 0.20 to 0.40 mm)

Type III: Industrial gloves (thickness: 0.36 to 0.85 mm)
Type IV: Special gloves

They can also by classified by weight and thickness:[2]

Ultra/very light weight: <0.20 mm
Light weight: 0.20 to 0.31 mm
Medium weight: 0.31 to 0.46 mm
Heavy weight: >0.46 mm

As can be seen there are no sharp borderlines between the different types of gloves.

Disposable gloves can be of different shapes, usually more or less transparent, with a pale white or beige color. Some gloves are extremely thin, made of polymeric film material, sometimes mounted on paper, and available in both sterile and non-sterile varieties. The most common disposable gloves are those manufactured by the dipping procedure described below. In recent years gloves for special purposes have been developed, e.g., gloves with increased thickness on the fingertips designed for handling cytostatic agents (Chem Plus Glove, Chemo Safety® Systems, San Diego, U.S. and Sempermed® Protector, Semperit, Austria) and gloves made of polymeric materials with low allergenic properties (Elastyren®, Danpren Glove, Denmark and Tactylon™**, Tactyl Technologies, Inc., U.S.) aimed especially at those who have been sensitized to latex gloves.

Household gloves or domestic gloves are usually of a thicker quality, quite often with a velourized inside in order to minimize uncomfort due to hand sweating. They are usually made from natural rubber, PVC, or plastic impregnated textile.

Industrial gloves are of a heavier quality and both unsupported and supported gloves are available. They are usually made from natural or synthetic rubber materials, leather, textile, and combinations of these materials.

Special gloves are often gloves designed for a special kind of work or an occupational group, e.g., firefighters, butchers, divers, and electricians, and gloves used together with total encapsulating chemical protective suits. They can also have a special design, e.g., with cuffs, extra long sleeves, strengthening of fibers, or materials with special quality.

II. MATERIALS

Protective gloves can be made of rubber and plastic materials, leather, textile, and combinations of these materials.[1-3] The quality can differ due to manufacturing processes and additives. A survey of the materials is presented in Tables 1 and 2.

---

*Sempermed® is a trademark of Semperit, Austria.
**Tactylon™ is a trademark of Tactyl Technologies, Inc.

TABLE 1 Survey of Rubber (Natural and Synthetic) Materials Used for Protective Gloves

Material name	Trade name	Chemical composition
Natural rubber		*cis*-Isoprene
Synthetic rubber materials		
Butyl rubber		Isobutene/isoprene
Chloroprene	Neoprene®	Polychloroprene
Fluor rubber	Viton®	Vinylidene fluoride /hexafluoropropene
Nitrile rubber		Acrylic nitrile/butadiene
Styrene-butadiene	Elastyren®	Styrene-butadiene
Styrene-ethylene-butadiene	Tactylon™	Styrene-ethylene-butadiene

Note: Combinations of polymers are also available as natural rubber/chloroprene, fluor rubber/chloroprene.

TABLE 2 Survey of Plastic Polymeric Materials Used for Protective Gloves

Material name	Chemical composition
EMA	Ethylene-methylmethacrylate
EVOH	Ethylene-vinylalcohol
Polyethylene (PE)	Polyethylene
Polyvinylalcohol (PVA)	Vinylalcohol
Polyvinylchloride (PVC)	Vinylchloride
PE/EVOH/PE	PE and EVOH laminate

A. Rubber Materials (Natural and Synthetic)

Natural rubber (NR) — Gloves are made from natural latex (90 to 95%) and several additives (5 to 10%), some of which can cause allergic reactions both in contact with the skin during the manufacturing process and from the final products. In order to reduce the risk of skin contact during weighing and processing operations, the rubber chemicals are available as master batches in preparations such as flake, pearls, surface-coated powders, or wax granulate, free from dust and providing a clean handling. Natural rubber is used for all types of gloves — thin sterile surgical gloves, domestic gloves, as well as heavy supported industrial gloves. Natural rubber has a very high elasticity compared to other glove materials used for manufacturing protective gloves.

Butyl rubber (IIR) — A copolymer of isobutene (97 to 99.5%) and isoprene (0.5 to 3%), butyl rubber has good resistance to light and ozone, low air permeability, excellent flexing properties and heat resistance, good flexibility at low temperature, good tensile strength and tear resistance. It does exhibit very poor resistance to petroleum oils and gasoline but excellent resistance to corrosive chemicals, vegetable oils, phosphate ester oils, and some ketones. Gloves for industrial use are available.

Chloroprene rubber (CR) — Neoprene® (DuPont de Neumours & Company), a polychloroprene, was the first commercial synthetic rubber, invented by E. I. DuPont de Neumours & Company in 1930. Moderately oil resistant, with good weather and ozone resistant properties, Neoprene® is used in all types of gloves such as thin surgical gloves, household gloves, supported and unsupported industrial gloves, and also in combination with natural rubber and/or textile.

Fluorocarbon rubber (FPR) — Viton®* (DuPont de Neumours & Company) is a copolymer of vinylidene fluoride and hexafluorpropylene. Unsupported gloves are available with good resistance against a wide variety of chemicals, especially most chlorinated and aromatic solvents, and have extreme heat and oil resistance. They are very expensive compared to gloves of natural rubber.

Nitrile rubber (NBR) — Nitrile butadiene rubber is a copolymer of acrylonitirile and butadiene, available as household and unsupported industrial gloves. These gloves have considerable resistance to oils, fuels, and certain solvents.

Styrene-butadiene rubber (SBR) — Copolymer of styrene and butadiene with a certain chain length. The only kind of glove available today in this material is a sterile surgeon's disposable glove (Elastyren®, Danpren Glove, Denmark). These gloves do not have the same elasticity as latex gloves but can be classified as hypoallergenic.

Styrene-ethylene-butadiene rubber — Tactylon™ is a copolymer of styrene, ethylene, and butadiene. In this material there are thin, disposable, sterile surgeon's gloves and non-sterile examination gloves available. They are also described as hypoallergenic.[4]

B. Plastic Polymer Materials

Ethylene-methylmethacrylate (EMA) — A copolymer of ethylene and methylmethacrylate. Foils of this material are used for punching/welding thin, transparent, disposable gloves. They are mounted on paper; sterile and non-sterile gloves are available.

Polyethylene (PE) — Polyethylene gloves are available as thin, disposable gloves. They are manufactured by punching/welding thin foils of PE and are sometimes mounted on paper. Gloves of different thicknesses and with patterned surfaces are available as well as with extra long sleeves (e.g., veterinarian use). They have a rather wide application in hospital work, food handling, work with curing-plastic materials, painting, and handling electronic components. The protective effect is more dependent on the strength of the seams than the chemical resistance of the material itself.

---

* Neoprene® and Viton® are registered trademarks of DuPont.

Polyvinylalcohol (PVA, PVAL) — Is a polymer of vinylalcohol. It is available as industrial gloves with good resistance to organic chemicals but not resistant to water and water-based solutions.

Polyvinylchloride (PVC) — Vinyl (chloride) plastic is a polymer of vinyl chloride. All types of gloves are available. The protective effect is good against water and most aqueous solutions, detergents, and diluted bases and acids. It has only restricted resistance to organic solvents.

PE/EVOH/PE — Is a laminate of polyethylene-ethylenevinylalcohol copolymer-polyethylene. It is available in two gloves manufactured by punching/welding of the silver-colored foil-laminate (Silver Shield™, Siebe North, Inc., U.S. and 4H™-glove*, Safety 4, A/S, Denmark). These gloves have excellent resistance against a wide range of chemicals, but their fitness for use in certain situations can be restricted due to the lesser elasticity of the material.[5,6]

III. MANUFACTURING

The gloves can be manufactured by a dipping procedure, by punching and welding of plastic film sheets, and by sewing. Gloves with linings are manufactured by a combination of the sewing and dipping procedures. Most rubber and plastic gloves are manufactured by the dipping method.

A. Gloves Made by Dipping

The dipping process of natural rubber latex has been thoroughly described by Gorton, Pendle, and Stern.[7-9] Molds or formers made of porcelain or metal in the shape of hands of different sizes are mounted on a rack and are slowly dipped into a solution or suspension of the polymeric material, natural latex, synthetic rubber, or plastic polymer. The thickness of the gloves depends on how many times the molds are dipped (Figures 1 and 2).

1. Compounding of Rubber Latex

Fresh latex from the rubber tree *Hevea brasiliensis* contains about 36% rubber (*cis*-isoprene), 60% water, 1.7% resins, 2% proteins, ash, and sugars. The latex suspension has to be preserved and concentrated; otherwise, coagulation takes place. Usually ammonia is used as a preservative, and it must be added at the earliest possible moment. High-ammonia (HA) latex contains about 0.7% ammonia. Low-ammonia (LA) latex contains about 0.2% ammonia and additional preservatives for example, sodium pentachlorophenate (LA-SPP), zinc oxide, tetramethylthiuram disulfide (LA-TZ), and boric acid (LA-BA) are usually added. All of them are suitable for use in the dipping process. The latex suspension also has to be concentrated to a rubber content of about 60% by

---

* 4H™-Glove is trademark of Safety 4®.

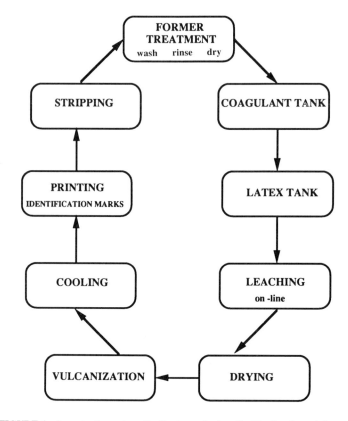

FIGURE 1 Layout of an automatic dipping unit, described by Pendle and Gorton.[8]

centrifugation, creaming, or evaporation before the dipping process. The high-ammonia latex is usually deammoniated to 0.2 to 0.3% ammonia content by evaporation or by adding a calculated amount of formaldehyde solution before the preparation of the latex dipping mixture.

2. Vulcanizing Systems

In natural rubber latex dispersions, sulfur is the primary vulcanizing agent, with zinc oxide as an activator and the addition of one or more accelerators. Vulcanization is an irreversible process where the polymer chains are crosslinked and the material becomes elastic. Certain chemicals are used to influence the rate of the vulcanization process. Ultra-fast accelerators commonly used in the manufacturing of gloves are zinc diethyldithiocarbamate (ZDC) or zinc dibutyldithiocarbamate (ZDBC) together with zinc 2-mercaptobenzothiazole (ZMBT). Vulcanizing can also be performed by adding sulfur donors, such as tetramethylthiuram disulphide (TMTD), and the activators thiourea or diphenylthiourea to the latex mix. Prevulcanized latex is commonly used in

Gloves: Types, Materials, and Manufacturing

FIGURE 2 Automatic glove dipping process.

dipping processes. The latex suspension is then mixed with a dispersion of sulfur, zinc oxide, and an ultra-fast accelerator to a temperature of approximately 70°C for about 2 h. Vulcanization after dipping is not necessary. The only compounding needed for prevulcanized latex is the adding of antioxidants.

Products of natural rubber latex are susceptible to oxidation and exposure to ozone and light. In the dipping processes, phenolic antioxidants (styrenated and hindered phenols) are commonly used for surgical gloves, examination gloves, domestic gloves, and condoms as they are nonstaining. Other antioxidant chemicals, such as phenylenediamines and substituted naphtylamines, stain the end products and are not commonly used in the manufacturing of protective gloves but can be used in dark-colored industrial gloves.

Sunlight can cause the latex rubber glove surface to resinify. The addition of pigments and/or UV-absorbers can improve the resistance against light. Protection to ozone can be achieved by adding so-called antiozonants.

TABLE 3 The Most Commonly Used Vulcanizing Agents, Activators, and Accelerators in Rubber Gloves

Function/ chemical name	Abbreviation	Typical dosage parts phr[a]
Vulcanization agents		0.3–1.5
Sulfur		
Stearic acid		
Tetramethylthiuram disulphide	TMTD	
Activators		0.1–2.0
Zinc oxide	ZnO	
Thiourea		
Diphenylthiourea		
Accelerators		0.3–1.5
Ultra-fast accelerators (cures at 100°C, 30 min)		
Zinc diethyldithiocarbamate	ZDC	
Zinc dibutyldithiocarbamate	ZDBC	
Medium-fast accelerator (cures at 100°C, 40 min)		
Zinc 2-mercaptobenzothiazole	ZMBT	

[a] Parts phr = parts per hundred of rubber, by dry weight.

J. Taylor has listed in tables a considerable number of different kinds of chemicals that are used in the rubber and plastic industry, giving the chemical names, trade names, and suppliers.[10] Information about the use, safe handling, and toxicity is also available in handbooks of different kinds.[11-14] The chemicals most commonly used in manufacturing of rubber and plastic gloves by the dipping process are presented in Tables 3 through 5.

Adverse effects from the latex sap, from the rubber and plastic additive chemicals used in the manufacturing of gloves, and from the use of gloves will be described in more detail in other chapters in this book.

3. Dipping Processes

Water-soluble ingredients are added to the latex as aqueous solutions, and water-insoluble ingredients are added as emulsions. Sulfur, zinc oxide, accelerators, and other solid ingredients are prepared in a ball mill. When preparing dispersions, it is essential that the particle size of the material is less than 5 μm. Particles that are too large will cause problems, such as settling in the dipping tanks, and can cause imperfections on the surface of the rubber product. It is also important that the different ingredients are added to the latex in correct order in order to maintain a stable mixture. When the mixing is completed, the latex mix is usually stored (matured) for about 16 h at room temperature with gentle stirring continuously. After storing, the latex mix is cooled to about 18 to 20°C and strained through an 80 to 100 mesh gauze. The latex mix is then fed to the dipping tanks.

TABLE 4 The Most Commonly Used Antidegradants and Stabilizing Agents in Rubber Gloves

Function/chemical name	Typical dosage parts phr[a]
Antioxidants	0.5–2.0
Styrenated phenols, non-staining	
2,4-dimethyl-6-(methylcyclohexyl)-phenol (Permanax WSL)	
Hindered phenols, non-staining	
Xylenol-aldehyde condensation product (Wingstay)	
2,2-methylene-*bis*(4-methyl-6-*t*-butylphenol) (BPH)	
Substituted naphtylamines, staining	
Phenyl-β-naphtylamine (PBN)	
Phenyl-α-naphtylamine (PAN)	
Phenylenediamines, staining	
Isopropyl-phenyl-phenylenediamine (IPPD)	
N,N'-bis(1,4-dimethylpentyl)-*p*-phenylendiamine (77PD)	
N,N'-bis(1-ethyl-3-methylpentyl)-*p*-phenylendiamine (DOPD)	
Stabilizers	0.2–0.5
Potassium caprylate	
Potassium laureate	
Potassium hydroxide	
Alkylphenol/ethyleneoxide	

[a] Parts phr = parts per hundred of rubber, by dry weight.

TABLE 5 The Most Commonly Used Heat-Sensitive Agents and Coagulants Used in the Dipping Process of Latex Rubber Gloves

Function/chemical name	Typical dosage
Heat-sensitive agents	
Polyvinyl methylether	10% solution/7.5–10 parts by weight
Polypropylene glycols	25% solution/10 parts by weight
Zinc oxide	50% dispersion/1–2 parts by weight
Ammonium acetate	15% solution/10 parts by weight
Coagulants	
Calcium nitrate	10–50% in water and/or methylated spirit
Cyclohexylamine acetate	
Zinc chloride	

There are three basic variations of dipping processes.

Straight dipping — The former is immersed in the latex mix, withdrawn very slowly, and the latex adhered to the former is then dried and vulcanized. A greater thickness can be obtained by drying the first layer and then reimmersing the former in the latex. This method is used for production of thin-walled products, such as condoms. The thickness of the rubber layer is only 0.05 mm. The thickness of the rubber layer depends on the total content of solids and the viscosity of the latex mix.

Coagulant dipping — The former is immersed in a coagulant solution, withdrawn, partly dried, and then immersed in the latex dispersion. After a predetermined time, the former is withdrawn, dried, and vulcanized. Coagulant solutions of calcium nitrate or zinc chloride (10 to 50%) in water and methylated spirits are commonly used. The coagulant solution sometimes contains a lubrication agent in order to facilitate the stripping from the formers. After one coagulant dip, the thickness of the deposit obtained will be around 0.2 to 0.8 mm. The thickness is dependent on the nature and concentration of the coagulant used, the dwell time, the total content of solids in the latex mix, and the viscosity. This is the most common process for manufacturing all kinds of rubber gloves, usually on highly automated production lines.

Heat-sensitive dipping — The formers are heated to 50 to 80°C and then dipped in the same way as for straight dipping. The thickness after one single heat-sensitive dip can be up to 4 mm and is dependent on the characteristics of the mix, the temperature and heat capacity of the former, and the dwell time in the latex mix. This process is used for thick-wall products such as electricians' gloves.

4. Leaching, Chlorination, and Lubrication

After the last dip the process continues with leaching, drying, and curing procedures. Other additional treatments that may be necessary, either as a part of the production process or separately, include leaching, chlorination, and lubrication. Dipped products are usually leached, which means that they are washed in water to remove water-soluble materials, thus lowering the water absorption of the material and improving the electrical resistance of the product. Leaching is particularly important for products like teats and medical devices made of latex, in order to reduce the content of extractable latex proteins which are known to cause hypersensitivity of Type I. It is also vital where maximum electrical resistance is needed, e.g., for electricians' gloves. The leaching process can be carried out either as on-line or off-line operations.

On-line leaching is carried out before drying. Hot water (60 to 80°C) should be used, since it gives a more rapid extraction of hydrophilic materials than cold water. As the time for on-line leaching is restricted in highly automated production lines, there will be no complete removal of water-soluble materials. It is only used on products for which the extraction of water-soluble substances is not critical.

Off-line leaching is carried out on the dried, vulcanized film and is a slow process. The extraction can take hours or days, depending on the thickness and the properties required for the product. This is a separate operation, and the long extraction does not interfere with the primary production process and hold up the production. It is always used where complete removal of water-soluble materials is essential, e.g., for electricians' gloves.

Chlorination is used to reduce the surface drag. The product is immersed in a dilute, aqueous chlorine solution (0.3%) for 2 to 5 min. The chlorine reacts with the rubber surface of the product, and this results in a lower coefficient of friction. The chlorination also results in a reduced level of extractable latex proteins, both due to the extra leaching achieved but also due to formation of insoluble forms of some proteins.[15] After chlorination it is necessary to wash the products to remove excess chlorine. This is done by using dilute, aqueous ammonia solution and then rinsing in water. The chlorination is permanent and is commonly used for products that are reused, like household gloves.

Lubrication is an alternative to chlorination to reduce the surface drag. Talc, cornstarch, lycopodium, and silicone oils can be used as lubricants, depending on the end use of the product. Powder lubricants are usually applied on-line, where the formers are dipped either in beds of powder or in water or alcoholic suspensions. Water-based powder suspensions have to be preserved. An irregular finish or a patterned surface on the glove is sometimes important to give a good grip in use. This can be achieved by using roughened formers so that, after the stripping procedure, the roughened finish is on the outside. Alternatively, solvent roughening is used to produce a pattern on the surface of the dipped gloves by immersing them in a rubber swelling solvent for a very short time, e.g., less than 10 s in toluene. This process is usually performed, after two coagulant dips, on the secondary layer of latex.

5. Drying, Curing, and Stripping

The dipped products are usually dried and vulcanized in an oven with hot air. The formers with the deposit are partly dried at a temperature of 80 to 90°C before the final curing stage, which is performed at temperatures between 100 to 140°C.

The final basic operation is the removal of the product from the former — stripping. In almost all cases, it is carried out manually but with the use of mechanical devices, such as compressed air nozzles. During the stripping the product is usually turned inside out. Before stripping the gloves from the formers, flocking and powdering processes may take place. In some cases, as, for example, with household gloves that have been flocked on the former, the stripping will result in the correct surface being on the outside.

6. Polymer Blends and Polymer Composites

In order to have improved physical properties of the gloves, such as tensile strength, elongation at break, tear strength, as well as excellent solvent resistance, it can be valuable to use combinations of different polymers. This can be achieved by using blends of the polymers in the latex dipping mix; e.g., natural rubber can be combined with nitrile rubber or polychloroprene rubber.

It is also possible to use the process of lamination dipping, where the bulk of the glove on the inside can be made of natural rubber, and the outer layer is formed by an additional dipping in a nitrile rubber or polychloroprene rubber mix.

The manufacturing of plastic gloves is performed in almost the same way as for the rubber latex gloves.[14] A plastic polymer mix, a plastisol or master batch, usually contains polymer, stabilizers, plasticizers, lubricants, and pigment. A plastisol is a dispersion of the polymer in a low-volatile plasticiser such as dioctyl phthalate. A organosol is a plastisol diluted with organic solvents like aromatic hydrocarbons or ketones. Organosols are used for coagulant dipping and plastisols for heat-sensitive dipping. Plastic polymer mixes also can be used for spreading on fabrics, which then can be used for manufacturing gloves by sewing.

B. Supported Gloves

Supported gloves are based on natural or synthetic latex-coated fabric liners, which can be made of knitted cotton or rayon or woven fabrics of cotton, wool, or synthetic blends. In the manufacturing of this kind of gloves, the fabric-covered former is dipped into the coagulant solution and then into the latex mix, dried, and vulcanized. The rubber layer only penetrates the outer part of the fabric and not through the complete fabric. There are two alternative methods of dipping fabric-lined gloves. In the first the viscosity of the latex mix is increased so that fabric penetration is controlled on immersion. After the first strait dipping, the thickness of the polymer layer can be built up by additional dips. Another method without excessive penetration of the fabric is to use hot formers and a heat-sensitive latex mix. The former is fitted with the liner and heated to 60 to 80°C and then dipped into the latex mix. The heating of the former sets up a temperature ingredient across the fabric which allows latex to penetrate only the surface before the heat causes the compound to gel. Natural latex or synthetic latex can be used separately or in blends for manufacturing supported gloves for heavy-duty industrial applications. Partly dipped gloves, where only the palms of the gloves are coated, can also be made.

C. Gloves Made by Sewing and Knitting

Gloves manufactured by sewing can be made of knitted fabric (cotton, nylon stretch, and other synthetic blends), of woven fabric (cotton, synthetic blends), of woven and/or knitted fabric, impregnated with natural latex or synthetic latex, and from leather (chromium or vegetable-tanned) and leather/woven fabric combinations. These kinds of working gloves have a wide range of applications. Thin, knitted gloves of cotton or nylon stretch can be used as inner gloves in order to reduce discomfort due to hand sweating and to avoid the risk for rubber contact allergy.

New kinds of knitted, cut and puncture resistant gloves have recently been made commercially available. They are made by knitting of special kinds of fibers, e.g., Repel™, cut-resistant glove liners made of Kevlar® and Lycra® fibers, LifeLiner™, stick- and cut-resistant gloves made of Medak™, Kevlar®, and Lycra® (DePuy and DuPont Orthopedics™), Perry® Cut-Resistant gloves made of Spectra® Fiber (Smith & Nephew Perry®), and Medarmor™ Cut- and

Puncture-resistant gloves made of Kevlar® with the fingertips having a puncture-resistant coating (Medical Armor Corporation).* They are all supposed to be used with sterile latex surgical gloves on top. The cut/puncture-resistant gloves may be reused about 10 to 12 times after washing, drycleaning, and sterilizing through autoclaving.

Gloves for work in cold areas have extra thick linings or synthetic fur linings, and gloves for work with hot objects are made of fabrics from heat-resistant fibers. Gloves made by sewing are of variable design and it is common that different materials are combined.

D. Gloves Made by Punching and Welding (Foil Gloves)

These kinds of gloves are made from two plastic polymer films (single-layered or laminates) punched out and welded in the same moment. They are manufactured in different sizes, but the fitness to the hand and the fingers is not comparable to gloves manufactured by the dipping procedure. The welded seams are usually the weak point of the glove, even if some of the polymer materials themselves have a good strength and excellent resistance to hazardous chemicals.

E. Summary

The protective effects of gloves against hazardous agents are among other factors dependent upon:

- Manufacturing process, material formulation, combinations, and material thickness
- Manufacturing process control
- Quality control of the end product
- Storing conditions (latex gloves are susceptible to UV light, X-rays, and ion radiation)
- Exposure to chemical mixtures and/or sequential exposures
- Mechanical strength

The side effects on the skin when using protective gloves are well known, and the most common reasons are

- Allergens in the glove materials such as
    Rubber chemicals and/or latex proteins in latex glove
    Preservatives in glove linings
    Chromium in leather gloves

---

* Repel™, LifeLiner™, and Medak™ are trademarks of DePuy-DuPont Orthopedics. Kevlar® and Lycra® are registered trademarks of DuPont® Perry® is a registered trademark of Smith & Nephew Inc. Spectra® Fiber is a registered trademark of Allied-Signal, Inc. Medarmor™ is a trademark of Medical Armor Corporation.

- Irritating agents such as glove powders (cornstarch or talc)
- Occlusive effect causing increased hand sweating
- Constitutional or acquired inclination to develop hand eczema

As manufacturing changes that are intended to decrease the latex protein content and/or decrease or replace rubber chemicals can significantly affect the safety and effectiveness of the gloves, it is, therefore, important to choose gloves that have been proven to fulfill appropriate quality control testing.

For people already sensitized to rubber chemicals and those with a latex protein allergy, there are "hypoallergenic" gloves produced by some manufacturers. The term "hypoallergenic" is allowed by the FDA to be used for labeling latex gloves that pass the Modified Human Draiz Test, but there are reports of allergic reactions in latex-sensitive individuals to "hypoallergenic" gloves.[16]

The term "hypoallergenic" has also been used by Heese et al. for thiuram-free gloves as a therapeutic alternative. They have prepared lists with "hypoallergenic" surgeon's and examination gloves that contain brand name of the glove, manufacturer, allergens (immediate/delayed), and irritants. [17]

Recently, Knudsen et al. presented a method for quantitative determination of thiuram and carbamate derivatives released from rubber gloves into synthetic sweat. The method is based on determination of ester derivatives using gas-chromatographic/mass spectrometry and isotope dilution. The accessibility of quantitative chemical analysis of thiuram and carbamate compounds used in glove manufacturing makes it possible to propose threshold values for release of these compounds in the near future.[18]

REFERENCES

1. Mellström, G. A., Protective gloves of polymeric materials: experimental permeation testing and clinical study of side effects, *Arbete och Hälsa*, National Institute of Occupational Health, Solna, Sweden, 1991.
2. Johnson, J. S. and Anderson, K. J., Eds., *Chemical Protective Clothing*, Vol. 1, AIHGA, Akron, OH, 1990.
3. Harper, C. A., Ed., *Handbook of Plastics and Elastomers*, McGraw-Hill, New York, 1983.
4. Lahti, A. and Turjanmaa, K., Prick and use tests with 6 gloves brands in patients with immediate allergy to rubber proteins, *Contact Derm.*, 26, 259, 1992.
5. *4H™ Chemical Protection List*, Safety 4, A/S, Lyngby, Denmark, 1991.
6. Tobler, M. and Freiburghaus, A. U., A glove with exceptional protective features minimizes the risk of working with hazardous chemicals, *Contact Derm.*, 26, 299, 1992.
7. Gorton, A. D. T., Natural rubber gloves for industrial use, *NR Technology*, 15, 7, 1984.
8. Pendle, T. D. and Gorton, A. T., Dipping with natural rubber latex, *NR Technical Bulletins*, The Malaysian Rubber Producers' Research Association, Hertford, England, 1980.

9. Stern, H. J., *Rubber: Natural and Synthetic.*, 2nd ed., Maclaren and Sons, Ltd., London, 1967, chap. 10.
10. Taylor, J. S., Rubber, in *Contact Dermatitis,* 3rd ed., Fischer, A. A., Ed., Lea & Febiger, Philadelphia, 1986, chap. 36.
11. *The Rubber Technologists' Pocket Book,* Vulnax International Limited, Vulnax Scandinavia AB, Göteborg, Sweden.
12. *Toxicity and Handling of Rubber Chemicals,* 2nd ed., British Rubber Manufacturers' Association, Ltd., Code of Practice, Birmingham, 1985.
13. Nutt, A. R., *Toxic Hazards Rubber,* Elsevier Applied Science Publishers, London, 1984.
14. Gächter, R. and Müller, H., Eds., *Plastic Additives Handbook,* 2nd ed., Hanser Publishers, Munich, 1984.
15. Subramaniam, A., Reduction of extractable protein content in latex products, Sensitivity to latex in medical devices, in Proc. Int. Latex Conf., Baltimore, November, 1992, Food and Drug Administration, 1992, 63.
16. Andersen, F. A., ODE perspective on the labeling of latex-containing medical devices, Sensitivity to latex in medical devices, in Proc. Int. Latex Conf., Baltimore, November, 1992, Food and Drug Administration, 1992, 69.
17. Heese, A., von Hintzenstern, J., Peters, K.-P., Koch, H. O., and Hornstein, O. P., Allergic and irritant reactions to rubber gloves in medical health services, *J. Am. Acad. Dermatol.,* 25, 833, 1991.
18. Knudsen, B. B., Larsen, E., Egsgaard, H., and Menné, T., Release of thiurams and carbamates from rubber gloves, *Contact Derm.,* 28, 63, 1993.

PART II

RULES AND REGULATIONS

4

European Standards on Protective Gloves

Gunh A. Mellström and Birgitta Carlsson

TABLE OF CONTENTS

I. Introduction 39
II. European Directives 40
III. Standards Related to Protective Gloves 42
References 43

I. INTRODUCTION

The decision of the European Economic Community (EEC) to create a single European market by the end of 1992 meant among other things, a free movement of goods. All the EEC countries have their own laws on product safety, and this can cause technical barriers to trade. In order to eliminate this problem for business and take into account the recommended high level of worker protection, the European Community Ministers agreed on a "New Approach to Technical Harmonization and Standards" in May 1985. This resulted in a Council Directive (Law) in the field of Personal Protective Equipment adopted in 1989, containing essential safety requirements. The technical details were left to the European standardization organizations.[1]

The body responsible for establishing the new standards for Europe is the Committee for European Normalization (CEN). In CEN there are representatives from all the EEC countries but also from the EFTA (European Free Trade Association) countries. The Technical Committee, CEN/TC 162 "Protective Clothing Including Hand and Arm Protection and Lifejackets" started working in 1989 and is divided into nine working groups:

Protective clothing (general)
Protective clothing against heat and fire

Protective clothing against chemicals
Protective clothing against foul, wind, and cold
Protective clothing against mechanical impact
Lifejackets
Various
Gloves
Motorcycle riders' protective clothing

The standardization procedure in CEN can follow (1) a questionary procedure (PQ) where, e.g., an ISO standard is available and sent out to CEN members (for 2 months); a revised document is then circulated for voting or (2) a technical procedure where a draft standard is prepared in a TC, sent out to CEN members for 6 months enquiry, and, after the comments have been considered, the revised document is circulated for voting (2 months). When a European Standard has been accepted, it will then be implemented by the CEN members within 6 months as a national standard. The standardization procedure is a continuous process, and at present some standards exist only in draft form (prENs), but they will lose their pr prefix once they have been approved by all member states.[2,3]

The work on international standardization in the field of protective clothing was initiated in 1964 in the International Standards Organization (ISO), Technical Committee 94 (ISO/TC 94) "Personal Safety — Protective Clothing and Equipment". The Sub-Committee 13 (Protective Clothing) of ISO/TC 94 was formed in 1981 and consists of five working groups. The working methods, results, and cooperation between the two standardizing organizations ISO and CEN and also the standardizing organization, the American Society of Testing and Materials (ASTM) in the U.S., have been described by T. Zimmerli.[4]

II. EUROPEAN DIRECTIVES

In the field of Personal Protective Equipment (PPE), two complementary Directives were adopted in 1989 and was due to enter into force in 1992. PPE is defined as any device or appliance designed to be worn or held by an individual for protection against one or more safety and health hazards in the execution of the user's activity.[1,5]

1. The Council Directive on the approximation of the laws of the Member States relative to the design of personal protective equipment (89/686/EEC).[6] This directive includes PPE for both professional and nonprofessional use. It defines, in particular, the certification procedures.
2. The Council Directive concerning the minimum safety and health requirements for the use by workers of PPE at the workplace [(89/656/EEC), third

individual directive within the meaning of Article 16(1) of Directive 89/391/EEC]. This Directive essentially defines the employer's obligations.

The assessment procedures of PPE relates to the control of the design and the control of their production. The target is to give the users the assurance that the device on the market fulfills the requirements. Application for EC-type examination is to be made by the manufacturer or his authorized representative. The authorized representative is to be established in the Community. Products meeting the requirements are to carry the CE mark, which implies that they can be marketed anywhere in the European Community countries.[1,5]

PPE categories according to the type and the risk:[6]

Category I — PPE of simple design, against minimal risks, which, when the effects are gradual, can be safely identified by the user in good time.

Category II — This category includes all types of PPE that are not included in categories I and III.

Category III — PPE of complex design, to protect against mortal danger or against dangers that may seriously and irreversibly harm the health. The user cannot identify the hazards in sufficient time.

Gloves are usually classified as types I or II.

Directive requirements dependent on the type of glove:

Type I. Gloves of simple design — for minimal risk application
    Conform with basic health and safety requirements
    Technical documentation file
    Declaration of conformity (requirements as set out in prEN420)
    Affix CE mark

Type II. Gloves of intermediate design (not simple nor complex design) — for intermediate risk
    Conform with harmonized European Standard or other verified technical specification
    Technical documentation file
    EC-type examination, tested by approved laboratories
    EC declaration of product conformity
    Affix CE mark on conformity, issued by an approved notified body

Type III. Gloves of complex design — for irreversible/mortal risks
    Conform with harmonized European Standard or other verified technical specification
    Technical documentation file
    EC-type examination, tested by approved laboratories and certified by approved notified bodies
    EC declaration of product conformity
    Affix CE mark
    Manufactured under a formal EC quality assurance system

III. STANDARDS RELATED TO PROTECTIVE GLOVES

Examples of EN Standards for protective gloves of different types and uses are listed in Tables 1 and 2. The standard (prEN 420), "General requirements for gloves," defines requirements for most kinds of protective gloves. Key points are fitness of purpose, innocuousness, sound construction, storage, sizing, measure of glove-hand dexterity, product information, and labeling.[7]

The ECC rules and regulations and CEN standards and their applications have been presented in several publications during the last few years.[1-5,7] A comparison of U.S. and CEN performance standards has been presented by Stull.[8]

TABLE 1 List of European Standards (EN) for Protective Gloves Prepared by CEN/TC 162

Document number	Title
EN 368	Protective Clothing for Use against Liquid Chemicals — Test Method: Resistance of Materials to Penetration by Liquids
EN 369	Protective Clothing for Use against Liquid Chemicals — Test Method: Resistance of Materials to Permeation by Liquids
prEN 374-1	Protective Gloves against Chemicals and Micro-organisms — Part 1: Terminology and Performance Requirements
prEN 374-2	Protective Gloves against Chemicals and Micro-organisms — Part 2: Determination of Resistance to Penetration
prEN 374-3	Protective Gloves against Chemicals and Micro-organisms — Part 3: Determination of Resistance to Permeation by Chemicals
prEN 388	Protective Gloves — Mechanical Test Methods and Specifications
prEN 407	Protective Gloves against Thermal Risks
CEN/TC162, N224, N 212	Gloves against the Effect of Vibrations
prEN 659	Protective Gloves for Firefighters
prEN 511	Protective Gloves against Cold
prEN 420	General Requirements for Gloves
prEN 421	Protective Gloves against Ionizing Radiation and Radioactive Contamination
prEN 464	Protective Clothing — Protection against Liquid Chemicals — Gas Leak Test

Note: CEN/TC 162: Protective clothing, including hand and arm protection and lifejackets; prEN: draft European standard.

TABLE 2 List of European Standards (EN) for Protective Gloves Prepared by CEN/TC 205

Document number	Title
prEN 455-1	Medical Gloves for Single Use — Part 1: Requirements and Testing for Freedom from Holes
prEN 455-2	Medical Gloves for Single Use — Part 2: Requirements and Testing for Physical Properties

Note: CEN/TC 205: medical devices; prEN: draft European standard.

REFERENCES

1. The Single Market. Personal Protective Equipment, DTI, Department of Trade and Industry and Central Office of Information, U.K., 1991.
2. Heffels, P. W. and Ziegenfuss, B. G., European standardization of protective clothing, in *Performance of Protecting Clothing: Fourth Volume, ASTM STP 1133,* McBriarty, J. P. and Henry, N. W., III, Eds., American Society for Testing and Materials, Philadelphia, 1992, 1001.
3. Ziegenfuss, B. G., European standardization of protective clothing, *4th Scand. Symp. Protective Clothing Against Chemicals and Other Health Risks,* Kittilä, Finland, Mäkinen, H., Ed., February 1992, NOKOBETEF, c/o HAZPREVENT, DK-1420 CopenhagenK, Denmark, NOKOBETEF IV, 1992, 49 (in English).
4. Zimmerli, T., International standards on protective clothing in ISO and CEN: cooperation or competition?, *4th Scand. Symp. Protective Clothing Against Chemicals and Other Health Risks,* Kittilä, Finland, Mäkinen, H., Ed., February 1992, NOKOBETEF, c/o HAZPREVENT, DK-1420 CopenhagenK, Denmark, NOKOBETEF IV, 1992, 56 (in English).
5. Mayer, A., European directives and standards relating to personal protective equipment, *Performance of Protecting Clothing: Fourth Volume, ASTM STP 1133,* McBriarty, J. P. and Henry, N. W., III, Eds., American Society for Testing and Materials, Philadelphia, 1992, 990.
6. Anon., Council Directive of 21 December 1989 on the approximation of the law of the Member States relating to personal protective equipment (89/686/EEC), *Official Journal of the European Communities* No. L 399/18, 30.12.89.
7. Ansell Edmont, *Guide to the New E.N. Standards on Hand Protection,* Ansell Edmont Europe N.V., Belgium, 1993.
8. Stull J. O., A comparison of U.S. and CEN performance standards for chemical protective clothing, *4th Scand. Symp. Protective Clothing Against Chemicals and Other Health Risks,* Kittilä, Finland, Mäkinen, H., Ed., February 1992, NOKOBETEF, c/o HAZPREVENT, DK-1420 CopenhagenK, Denmark, NOKOBETEF IV, 1992, 61 (in English).

5

Protective Gloves for Occupational Use — U.S. Rules, Regulations, and Standards

Norman W. Henry, III

TABLE OF CONTENTS

I. Introduction 45
II. Rules, Regulations, and Standards 46
III. Summary and Conclusions 48
References 50

I. INTRODUCTION

Gloves are used to protect one of man's most valuable tools — his hands. Our hands can do marvelous things and perform many tasks. These tasks may require the use of gloves for protection against exposure to hazardous physical, chemical, and biological agents during the normal work day. Gloves act as a barrier between our skin and the hazard encountered. Skin is a natural barrier of living tissue sensitive to physical effects, chemical absorption, and biological penetration. Once this natural barrier is broken our body is susceptible to harm and injury from any one of the three hazards. Gloves provide warmth from cold, insulation from heat, and resistance to sharp objects, chemicals, and biological organisms. They come in various colors, sizes, and shapes and are made of many different types of natural and synthetic materials. Despite their use and non-use, hand injuries continue to be one of the most frequently reported occupational injuries. These injuries are preventable if the correct gloves are selected and used for protection against the hazard. Selection of gloves for occupational use is the job of the safety professional. Guides, rules, and regulations for glove use have only recently become available to the safety professional through the efforts of various standard setting organizations that have developed performance standards for gloves. This chapter will discuss some of the current glove standards in the U.S.

II. RULES, REGULATIONS, AND STANDARDS

Glove standards can be divided into categories depending on the generic type of glove material used, type of work being done, and type of hazard encountered. For example, there are standards for gloves made of rubber, for gloves used by electrical workers, and for gloves used against heat and flame. Each standard was generated as a result of a need to evaluate performance of a material for specific workers against specific hazards. The majority of these standards in the U.S. have been drafted and written by voluntary consensus of standard-setting organizations such as the American Society for Testing and Materials (ASTM) and American National Standard Institute (ANSI). These standard-setting organizations have been the focal point for glove and other protective clothing standards. Various professional organizations representing their constituents have also adopted standards, such as the National Fire Protection Association (NFPA) and the American Dental Association (ADA). Federal standards for the testing of gloves also have been developed to meet military specifications and eliminate unnecessary or undesirable variations in the general sampling and testing procedures. International organizations such as the International Standard Organization (ISO) have also developed standards.

Most of the standards that have been adopted have been written to test the performance of gloves under various exposure conditions. The primary parameters evaluated have been physical strength, dexterity, abrasion, and heat and cold resistance. Resistance to swelling, degradation, permeation, and penetration are some of the more important chemical parameters evaluated, while biological resistance to liquids and microorganisms also have been evaluated. By measuring key effects in physical and chemical properties, such as tensile strength, thickness changes, and solubility, the performance of materials used to make gloves can be determined. Because much is known about the properties and behavior of natural and synthetic polymers and other materials of construction used for gloves, results of these performance tests conducted under actual use conditions can then be compared to original data on the generic materials to help predict resistance.

There is a broad range of test methods and test conditions. Standard test methods may involve a simple light test in a dark room to look for pinholes or imperfections in gloves, immersion of gloves in a chemical or biological liquid to see if penetration occurs, or more sophisticated permeation tests to determine breakthrough time and permeation rate. Some tests involve using the whole glove, while others may just require testing a sample or swatch of material. Exposure conditions can range from a splash test to complete liquid contact for varying lengths of time and at different temperatures. Equipment used for the tests also can vary in degree of complexity, detection sensitivity, level of sophistication, and cost: for example, the inflated-glove water-immersion test for determining pinhole imperfections vs. the use of an Instron for measuring tensile strength. Another criterion for these test methods is that they be reproducible and simple enough that most users can do testing for themselves at actual exposure conditions.

For the past 15 years one technical committee within ASTM, Committee F-23 on Protective Clothing, has been the focal point of voluntary consensus standards development for items of protective clothing, such as gloves. This committee charged with the responsibility to develop standard methods of test, terminology, classification, and performance specifications for clothing used to protect against occupational hazards, has generated 20 new standards on test methods, specifications, practices, and guides for physical, chemical, biological, and other hazards. These performance standards have set the pace for other standard setting organizations in the U.S. and internationally. The committee has sponsored four international symposia, published four standard technical publications (STPs), and compiled a single ASTM Protective Clothing reference on all ASTM protective clothing standards.[1] Standards on gloves are included in these publications.

One of the most notable standard methods developed within Committee F-23 was standard test method F739-91: *Test Method for Resistance of Protective Clothing Materials to Permeation by Liquids or Gases under Conditions of Continuous Contact.* It was the first standard method developed by F-23 and the one with the most impact on standards for gloves. The key parameters evaluated in this test method are glove breakthrough time and permeation rate. Because gloves and hands can come in direct contact with chemical hazards, it was recognized that they were susceptible to permeation and that some glove materials were more resistant than others. By measuring the chemical breakthrough time and subsequent permeation rate, one could determine chemical hold out (resistance) and protection capacity. This was particularly important for hazardous industrial solvents (benzene) and gases (ammonia). Upon acceptance of this standard, numerous permeation tests were conducted and glove permeation charts and guides were published. Glove users could at least now refer to how long a given glove will protect against exposure. Manufacturers, on the other hand, could work on improving glove performance by developing new products with better resistance. Because of the diversity of chemicals and the need for guidance on test method strategy a list of a standard battery of test chemicals was developed, F 1001-89: *Guide for the Selection of Chemicals to Evaluate Clothing Materials.* Now, most glove manufacturers publish the permeation results of their products using this list as a reference.

While chemical glove permeation standards were the first methods to be developed by Committee F-23 on Protective Clothing, another important area of standards development has been in the evolution of medical glove standards.[2] ASTM Committee D-11 on Rubber has been the focal point for this important standard-setting activity since the mid-1970s. Faced with the challenges of determining pinhole leaks in rubber surgeons' and examination gloves, this committee needed to develop better test methods than D 3577, *Specifications Rubber Surgical Gloves* and D 3578, *Specification for Rubber Examination Gloves* to address healthcare concerns about Acquired Immune Deficiency Syndrome (AIDS) and human immunodeficiency virus (HIV). After considerable review the committee published Standard Method D 5151: *Test Method for Detection of Holes in Medical Gloves* in 1990. This method resulted in an improved, more sensitive test to detect holes in medical gloves. While pinhole leaks were the primary concern in Committee D-11, Committee F-23 was focusing on penetration test methods for determining leaks with a synthetic blood test and a viral penetration test method using a bacteriophage surrogate to the AIDS and HIV viruses, Phi-X 174. These pass/fail test methods were evaluated this past year in interlaboratory round robin tests and currently are Emergency Standards ES 21: *Test Method for Resistance of Protective Clothing Materials to Synthetic Blood* and ES 22: *Test Method for Resistance of Protective Clothing Materials to Penetration by Blood-Borne Pathogens Using Viral Penetration as a Test System.*[3] Currently, these two test methods are being used to evaluate performance of clothing and glove materials, but improved methods specifically for gloves are being investigated and considered by Committee F-23.

III. SUMMARY AND CONCLUSIONS

A list of U.S. standards on gloves is shown in Table 1. These standard test methods were all developed by various standard-setting organizations such as ASTM, ANSI, NFPA, and ADA. They are performance standards used to evaluate gloves used for protection against various hazards in the work environment. These hazards include physical, chemical, and biological agents. The list is not complete, nor is there a description of the actual test method and conditions, since detailed procedures would consume much of this chapter. The list is intended to be a reference for identifying existing standards in the U.S.

As new technologies evolve the need for hand protection and improved glove test methods will continue. Standard-setting organizations in the U.S. and the rest of the world need to work together in developing these methods, since we now compete in a global market with no boundaries when it comes to protecting our fellow man. We should wear gloves that the job demands, select gloves that pass performance standards, and remember that many of the accomplishments of man would not be possible without protective clothing.

U.S. Rules, Regulations, and Standards

TABLE 1   U.S. Standard Methods for Gloves

Organization	Document number	Title
ASTM	C 852	Standard Guide for Design Criteria for Plutonium Glove Boxes
ASTM	D 120 E1	Standard Specification for Rubber Insulating Gloves
ASTM	D 3577	Standard Specification for Rubber Surgical Gloves
ASTM	D 3578	Standard Specification for Rubber Examination Gloves
ASTM	D 4115	Women's and Girl's Knitted and Woven Dress Glove Fabrics, Performance Specifications
ASTM	D 4679	Standard Specification for Rubber Household or Beautician Gloves
ASTM	D 5151	Standard Test Method for Detection of Holes in Medical Gloves
ASTM	D 5250	Standard Specification for Polyvinyl Chloride Gloves for Medical Application
ASTM	F 496	Standard Specification for the In-Service Care of Insulating Gloves and Mittens
ASTM	F 696	Standard Specification for Leather Protectors for Rubber Insulating Gloves and Mittens
ASTM	F 739-91	Test Method for Resistance of Protective Clothing Materials to Permeation by Liquids or Gases under Conditions of Continuous Contact
ASTM	F 903-90	Test Method for Resistance of Protective Clothing Materials to Penetration by Liquids
ASTM	F 955-89	Test Method for Evaluating Heat Transfer Through Materials for Protective Clothing Upon Contact with Molten Substances
ASTM	F 1001-89	Guide for Selection of Chemicals to Evaluate Clothing Materials
ASTM	F 1060-87	Test Method for Thermal Protective Performance of Materials for Protective Clothing for Hot Surface Contact
ASTM	D 4108-87	Test Method for Thermal Protective Performance of Materials for Clothing by Open Flame
ASTM	F 1342-91	Test Method for Protective Clothing Resistance to Puncture
ASTM	F 1383-92	Test Method for Resistance of Protective Clothing Materials to Permeation by Liquids or Gases under Conditions of Intermittent Contact
ASTM	F 1407-92	Test Method for Resistance of Chemical Protective Clothing Materials to Liquid Permeation — Permeation Cup Method
ASTM	ES 21	Test Method for Resistance of Protective Clothing Materials to Synthetic Blood
ASTM	ES 22	Test Method for Resistance of Protective Clothing Materials to Penetration by Blood-Borne Pathogens Using Viral Penetration as a Test System
ANSI/ADA[a]/ASA[b] S 3.40	No. 76	Nonsterile Latex Gloves for Dentistry, Guide for the Measurement and Evaluation of Gloves Which Are Used to Reduce Exposure to Vibration
ANSI/NFPA[c]		Standard on Gloves for Structural Fire Fighting
Federal standard	No. 601	Rubber: Sampling and Testing

[a]   ADA — American Dental Association.
[b]   ASA — Acoustical Society of America.
[c]   NFPA — National Fire Protection Association.

REFERENCES

1. ASTM Standards on Protective Clothing, Committee F-23 on Protective Clothing, 1990, American Society for Testing and Materials, Philadelphia.
2. Chatterton, J. R. and Culp, R. D., The evolution of medical glove standards, *ASTM Standardization News,* Aug. 1992, 26.
3. ASTM, Atmospheric analysis, occupational health and safety; protective clothing, in *Annual Book of ASTM Standards,* Sec. 11, Vol. 11.03, American Society for Testing and Materials, Philadelphia, 1993.

PART III

TESTING OF PROTECTIVE EFFECT

6

Testing of Protective Effect Against Liquid Chemicals

Gunh A. Mellström, Birgitta Carlsson, and Anders S. Boman

TABLE OF CONTENTS

I. Terms and Definitions 53
II. Degradation Testing 54
III. Permeation Testing 55
    A. Key Parameters 55
    B. Standard Test Methods 57
    C. Permeation Test Cells 58
    D. Factors Influencing the Permeation Test Results 60
    E. Evaluation of Test Results 63
        1. Flow Rate and Mixing 64
        2. Cell Size and Design 65
        3. Open or Closed Loop System 66
        4. ASTM vs. ISO/DIS Test Method 66
        5. Temperature 67
        6. Thickness and Formulation of the Material 67
    F. Intermittent Contact Permeation Testing 69
    G. Protection Index 70
    H. Modified Permeation Testing 70
IV. Penetration Testing 71
V. Summary 72
References 73

I. TERMS AND DEFINITIONS

Even if there are not always, or for all situations, regulations with demands on use or proven efficacy of protective gloves, there are several methods for testing the resistance to chemicals of polymeric materials used in chemical protective clothings and gloves.

Degradation is defined as a deleterious change in one or more physical properties of a protective clothing material due to contact with a chemical.[1]

Permeation is usually defined as the process by which a chemical migrates through the protective clothing material at a molecular level. The permeation process involves three stages: (1) sorption of molecules of the chemical into the outside surface of the material, (2) diffusion of the sorbed molecules through the material, and (3) desorption of the molecules from the inside surface of the material into the collecting medium.[1-3]

Penetration is usually defined as the flow of a chemical through closures, porous materials, seams and pinholes, or other imperfections in a protective clothing material on a nonmolecular level.[1]

Protective clothing material is defined as any material or combination of materials used in an item of clothing for the purpose of isolating parts of the body from direct contact with a potentially hazardous chemical.[1]

Protective glove material is defined as any material or combination of materials used in a glove for the purpose of isolating the hands and arms from direct contact with a chemical and/or microorganism.[2]

ASTM stands for the American Society for Testing and Materials.

BOHS stands for British Occupational Hygiene Society.

BSI stands for British Standard Institution.

CEN stands for Comité Européen de Normalisation.

CEN/TC 162 stands for CEN/Technical Committee 162: Protective Clothing, this committee has nine working groups (WG); WG 8: Gloves.

EN stands for European Standard (E), Norme Européen (F) and Europäische Norm (G).

ISO stands for the International Standard Organization.

prEN stands for draft European Standard.

II. DEGRADATION TESTING

The immersion test was one of the first tests used by the manufacturers in order to describe the chemical resistance of the protective gloves. Pieces of glove materials were immersed in various chemicals for a period of time and then visibly inspected and rated according to resistance, for example, as excellent, good, fair, or not recommended. The results can be misleading as more than one surface of the material is exposed to the chemical, i.e., the outside, inside, and cross sections. This will be a problem, especially when testing supported gloves.

The ability of the material to resist degradation by chemicals was previously commonly used method to evaluate the protective effect.[5] Several drafts of ASTM test methods for evaluation of protective clothing materials for resistance to degradation by liquid chemicals have been written. The test cell used is the ASTM permeation test cell, where the collecting side has been modified and only the outer surface of the glove is exposed to the chemical. The properties measured are thickness, weight, and elongation changes after exposure to the chemical. The degradation test method still needs some modifications before it can be considered a standard test method, but it can be used in the screening procedure when developing new polymeric materials for protective gloves or clothings.[4,6,7]

In the work program of CEN /TC 162 the determination of resistance to degradation is mentioned as a method to be developed.[2]

III. PERMEATION TESTING

During the last 10 years the use of permeation test methods has increased, and the first standard test method (ASTM-F 739) was promulgated by ASTM 1981, updated in 1985, and again in 1991.[1,8,9] In Europe, a British Occupational Hygiene Technology Committee working party was formed to develop a standardized test method for permeation of liquid chemicals through protective clothing materials.[10] This standard test method was submitted to the BSI and to the ISO and was presented as a Draft International Standard ISO/DIS, in 1988.[11]

In the European Committee for Standardization (CEN) work has been ongoing to establish EN concerning protective gloves against chemicals and microorganisms since 1992.[2,3,12,13] The promulgation of EN test methods for resistance of protective clothing material to permeation by chemicals will hopefully be the basis for efforts to create a reasonable system of rules and demands for the selection and use of protective gloves for working with hazardous chemicals.

A. Key Parameters

The principle of permeation testing is a flow-through system where a two compartment permeation cell of standard dimensions is used. The test specimen acts as an barrier between the first compartment of the cell, which contains the test chemical, and the second compartment through which a stream of the collecting medium (gas or liquid) is passed for the collection of diffused molecules of the test chemical or its component chemicals for analysis. The outside of the glove or clothing material continuously contacts the test chemical in excess and the collecting medium contacts the other (inner) side. The resistance is determined by measuring the breakthrough time and the permeation rate of the chemical through the test material. The definitions of the key parameters are similar in the different standard test methods, with some exceptions for the current version of ASTM F 739-91, where some new terms have been added — for example, breakthrough detection time, normalized detection time, minimum detectable permeation rate, and minimum detectable mass permeate.[9]

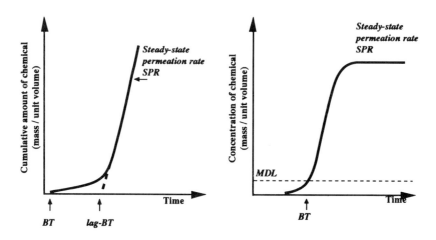

FIGURE 1 Permeation of chemicals through test specimen. Left: cumulative amount of chemical permeating vs. time (closed loop system). Right: concentration in collecting medium vs. time (open loop system). BT = measured breakthrough time, lag-BT = cumulative breakthrough time, MDL = minimum detectable level (concentration). (From Mellström, G. A., *Ann. Occup. Hyg.*, 35, 167, 1991. With permission of the British Occupational Hygiene Society.)

Breakthrough time (BT) is the elapsed time (in minutes) between the initial application of a test chemical to the appropriate surface of the material and its subsequent presence on the other side of the material, measured as described in the standards.

Breakthrough detection time (BDT) is the elapsed time (in minutes) measured from the start of the test to the sampling time that immediately precedes the sampling time at which the test chemical is first detected.

Time-lag breakthrough (lag-BT), sometimes called cumulative breakthrough, is described in the Draft International Standard 6529.4 ISO/DIS and is the extrapolation of the steady-state permeation portion of the cumulative permeation curve to time axis (in minutes)[11] (Figure 1).

Normalized breakthrough detection time (NBDT) is used in the ASTM F 739-91 and is defined, in an open-loop test, as the time (in minutes) at which the permeation rate reaches 0.1 $\mu g/cm^2/min$. In a closed-loop test, it is the time at which the mass of chemical permeated reaches 0.25 $\mu g/cm^2$.

Normalized BT (NBT) is used for the BT in minutes divided by the material thickness in millimeters. The normalized breakthrough has sometimes been used in investigational reports and makes it possible to do a direct comparison of test results between glove materials, independent of the thickness.[16,17]

Permeation rate (PR) is the mass of test chemical permeating the material per unit time per unit area and is usually measured in micrograms per minute per square centimeter, milligrams per minute per square meter, or milligrams per second per square meter (Figure 1).

Steady-state permeation (SP) is a state that is reached when the permeation rate becomes virtually constant and is usually measured in micrograms per minute per square centimeter, milligrams per minute per square meter, or milligrams per second per square meter (Figure 1).

Minimum detectable mass permeated is the smallest mass of permeant that is detectable with the complete permeation test system.

Minimum detectable permeation rate (MDPR) is the lowest rate of permeation that is measurable with the complete permeation test system.

Most chemical permeation testing has been performed according to ASTM F 739-85. This version is discussed below along with comments relative to the current edition, ASTM F 739-91, published in April 1992.

B. Standard Test Methods

When following a standard test method the permeation testing can be performed in different ways, as the protocols stated in the standard test methods may vary as may the test cell design. For permeation testing of chemicals there are two standard test methods, the ASTM F 739-91, last previous edition F 739-85, and the ASTM F 1383, and three draft standard methods, the ISO/DIS 6529.4, prEN 374-3, and prEN 369. The two prEN draft standards are supposed to be established as EN from December 31, 1992. There are some differences but also some similarities between those test methods as seen in the designations.

prEN 374-3: Protective Gloves Against Chemicals and Microorganisms — Part 3: Determination of Resistance to Permeation by Chemicals.

prEN 369: Protective Clothing For Use Against Liquid Chemicals; Resistance of Materials to Permeation by Liquids.

ISO/DIS 6529.4: Protective Clothing — Protection Against Liquid Chemicals — Determination of Resistance of Air-Impermeable Materials to Permeation by Liquids.

ASTM F 739-85: Standard Test Method For Resistance of Protective Clothing Materials to Permeation by Liquids or Gases.

ASTM F 739-91: Standard Test Method For Resistance of Protective Clothing Materials to Permeation by Liquids or Gases Under Conditions of Continuous Contact.

ASTM F 1383: Standard Test Method For Resistance of Protective Clothing Materials to Permeation by Liquids or Gases Under Conditions of Intermittent Contact.[14]

The purpose of prEN 374-3 is determination of permeation of protective gloves by solid or liquid chemicals where for the others it is the determination of protective clothing by liquid chemicals, and for ASTM F 739 and ASTM 1838, also by chemical gases under conditions of continued contact or intermittent contact, respectively.

The draft standard method prEN 369 is similar to the ISO/DIS 6529.4 and the prEN 374-3 to the ASTM F 739-85.

Variables in permeation testing that are of importance for the results are the nature of collecting the medium, flow rate of the collecting medium, volume of the collecting chamber, open or closed system, temperature, exposed area of the sample, sampling mode and frequency, method of analysis, etc.[6,15,19-24]

Collecting medium can be an inert gas such as dried air, nitrogen, helium, or a liquid. The liquid used as collecting medium must be proven not to influence the resistance of the test material. Water is the most commonly used liquid medium. The test chemical shall be freely soluble in the collecting gas or liquid.

Flow system can either be an open- or closed-loop flow-through system (Figure 2). In a closed-loop system the possibility of saturation should be considered. The assessment of the permeating chemical in the system can be continuous or by time-scheduled sampling, depending on the analytical technique and equipment used.

Temperature should be the most relevant to the use of the material, maintaining it constantly during the testing at ±1°C at the appropriate temperature is crucial.

Analytical system shall have sensitivity for the test chemical at levels of 1 mg to 0.1 µg/min/cm^2 minimum detectable permeation rate (MDPR).[2,3,9,11] According to the ASTM F 739-85 Standard Test Method, the minimum detectable level was not specified but in the current version, F 739-91, the system shall have a sensitivity of at least 0.1 µg/cm^2/min in open-loop testing, and a minimum sensitivity of 0.25 µg/cm^2 in a closed-loop testing system.[1,9]

Calculations and expressions of results. The calculation of the results is well described in the test methods as is the method of expression of the results. There are main differences in reporting the results — in the two prEN draft standard methods only the breakthrough time shall be reported, but in the ASTM F 739-85 standard method both BT and SPR, and in ASTM F 739-91 the BDT, NBDT, and SPR data shall be reported. In the ISO/DIS 6529.4 method the lag-BT and the mass permeated per square centimeter after 30 and 60 min shall be reported.

Specifications of the test conditions of ASTM, ISO/DIS 6529.4, prEN 369, and the prEN 374-3 methods are presented in Table 1.

C. Permeation Test Cells

The standard ASTM test cell is a glass (2 in. diameter) chemical permeation cell available from Pesce Lab Sales, Kennett Square, PA. There is also a smaller version of the ASTM test cell (1 in. diameter) available. The membrane is in a vertical position in contrast to most other cells where the glove membrane is in a horizontal position. Inlet and outlet of the collecting medium are situated almost at right angles to each other with the inlet directed at the test membrane (Figure 3). In the ASTM F 739 standard it is possible to use

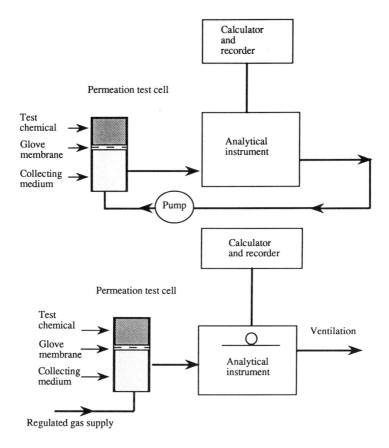

FIGURE 2 Flow system — closed loop system (top) and open loop system (bottom).

alternative test cells provided that they have been found equivalent to the standard reference cell (2 in. cell). A standard practice for the determination of equivalency of optional chemical permeation test cells to that of the ASTM cell has been developed, but still needs modification.[25]

The ISO/DIS test cell is a stainless steel permeation cell. The internal dimensions of the collecting compartment and its pipework are critical to the performance of the test.[11] In the ISO/DIS 6529.4 standard draft no alternative test cells are allowed. Drawings of the ASTM and ISO/DIS permeation test cells can be seen in Figure 3.

In the draft standard prEN 374-3 the ASTM standard test cell is described as the test cell that shall be used. However, in this draft standard, alternative test cells can be used if proven to be equivalent to the ASTM reference cell. In the draft standard prEN 369 the ISO/DIS test cell is described as the test cell that shall be used and no alternative cells are allowed. Specifications of the test cell data are presented in Table 2.

TABLE 1 Specifications of the Test Conditions for ASTM F 739, ISO/DIS 6529.4 and prEN Permeation Test Procedures

Test methods	ASTM F739 1985/1991	prEN 374-3	ISO/ DIS	prEN 369
System	Open or closed	Open or closed	Open	Open
Collecting medium flow rate				
Gas (open loop) Volc/min	5/0.5 – 1.5	5	30	30
Liquid (open loop) Volc/min	5/0.5 – 1.5	Sufficient	12	12
Liquid (closed loop) Volc/min	Sufficient	Sufficient	—	—
Reporting of results				
BT, min	BT/NBDT, BDT	BT	lag-BT	BT
Permeation	SPR/SPR	—	MASS	—
Detection limit				
Open loop, µg/cm^2/min	NS/0.1	1000	1	1
Closed loop	NS/0.25 µg/cm^2	1 mg/cm^2/min	—	—

Note: MASS = mg permeated/cm^2, 30 and 60 min after BT,
NS = not specified,
SPR = steady state permeation rate, µg/cm^2/min,
Volc/min = volume changes/min.

D. Factors Influencing the Permeation Test Results

Permeation is described as a molecular process by which chemicals are first adsorbed on the surface of the material, diffuse through the material along a concentration gradient, and are desorbed from the other surface of the material.[8] The mathematical theory of diffusion is based on Fick's first law and can be presented in the formula:

$$P = -D\partial C/\partial x$$

where:

P is the rate of transfer per unit area,
C is the concentration,
x is the distance within the material, e.g., thickness of the material, and
D is the diffusion coefficient, a proportionality constant which may be a function of chemical concentration.[24,26]

The diffusion coefficient, however, is usually treated as a constant. The theoretical mathematical model for the permeation process resulting in differential equations for the total amount of the chemical permeating per unit area at time, and for the permeation rate of the chemical as a function of time per unit area of the material, have been described and discussed by Schwope et al.[24] They pointed out that the detection of BT, which was defined as the time t at which the concentration of the chemical is detected in the collecting side of a permeation test cell, is dependent on the material thickness, sensitivity of the analytical equipment, and the cell size. Theoretical and experimental approaches have

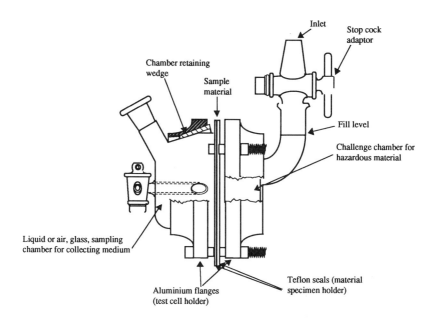

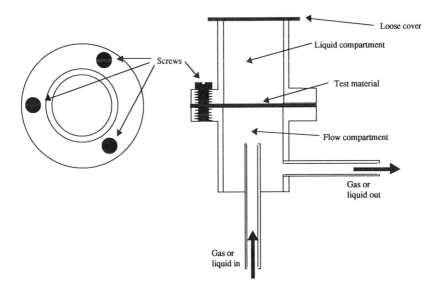

FIGURE 3 Top: ASTM (1-in.) and bottom: ISO/DIS permeation test cells. (From Mellström, G. A., *Ann. Occup. Hyg.*, 35, 153, 1991. With permission of the British Occupational Hygiene Society.)

TABLE 2 Test Cell Specifications

	ASTM F 739 prEN 374-3	ISO/DIS 6529.4 prEN 369
Permeation test cells		
Size and design of test cells	ASTM cell (2 in./1 in.)	ISO/DIS cell
Volume of collecting medium	100 ml/15 ml	17.2 ml
Exposed area	20.3 cm^2/5.3 cm^2	4.9 cm^2
Position of exposed area	Vertical/vertical	Horizontal
Ratio: area/volume	0.203/0.353	0.285
Alternative test cell allowed	Yes/yes	No

been taken to identify the parameters that have a crucial influence on the detection of the breakthrough time by Billing and Bentz.[26] Five different kinds (A through E) of permeation rate vs. time behavior due to interaction between the chemical and glove material have been described by Nelson et al.[27] The most typical pattern is the A type, where the permeation rate stabilizes at a "steady state" level. In the B type the permeation behavior is due to the material specimen being structurally modified by the test chemical, resulting in a change, where instead of stabilizing, the permeation rate gradually increases or decreases. In the C type a sudden very large increase in permeation rate occurs. The D type of behavior exhibits an initially high rate, decreasing slowly and eventually stabilizing during the exposure time, and occurs when there is moderate to heavy swelling of the material. The E type occurs when there is a high degree of swelling resulting in a continuous increase in the permeation rate with intermittent levels of steady state.[27] The manner and extent of effect from chemical interaction with the glove material can sometimes be hard to distinguish from the influence of different test conditions on the permeation test results.

The importance of how to present and evaluate test data and test conditions in a way that provides valuable information to the consumer has been discussed both by Jamke[28] and Kairys.[29] They pointed out that the most important variables to evaluate include the detector sensitivity, sampling method, system flow rate, surface of the sample, and the test system (open or closed) used.

From the formula given for calculating the permeation rate in an open-loop system it can easily be seen that the cell size, exposed area, and flow rate of the collecting medium in the test cell will have an influence on test results, the steady state permeation rate, as well as breakthrough time.[24]

$$P = (C \cdot F)/A$$

where:

P = steady-state permeation rate (PR) mg/m^2/min;
C = concentration of the chemical, mg/l;
F = flow rate of the collecting medium, l/min; and
A = exposed area, m^2.

The flow rate through the collecting medium compartment can also be expressed as volume changes per minute (volc/min).

Measurement of the concentration of the permeated chemical in the collecting medium is dependent on the detector sensitivity. A higher concentration is needed for a less sensitive detector. The breakthrough time (BT) will increase when the minimum detectable level increases. For the same glove material tested with different analytical equipment, different BT can be reported. In the ISO/DIS 6529.4 and prEN 369 draft standard methods the MDPR is specified to 1 $\mu g/cm^2/min$, and in prEN 374-3 to 1 $mg/cm^2/min$ and, thus, the BT is the time when PR = MDPR (Figure 4). In the ASTM F 739-85 a detection limit is not specified but will be reported. In the current edition, ASTM F 739-91, the detection limits are specified to 0.1 $\mu g/cm^2/min$ in open-loop testing systems and to 0.25 $\mu g/cm^2$ in closed-loop testing systems. The new term, normalized breakthrough detection time, NBDT, equals the time at which the permeation rate reaches 0.1 $\mu g/cm^2/min$ in an open-loop system and, respectively, when the mass of chemical reaches 0.25 $\mu g/cm^2$ in a closed-loop system.

When evaluating the measured BT values performed according to different test methods it has to be taken into account that the detection limits are specified at three different levels, 1 mg, 1 μg, and 0.1 $\mu g/cm^2/min$.

The concentration in the collecting medium is also dependent of the sampling method. The measurement can be either continuous or done by sampling following a time schedule. When the sampling frequency decreases from, e.g., every second minute to every five or ten minutes the BT will increase. Continuous measurement gives shorter BT values compared to measurement by sampling.[29]

The ratio between the surface area and the volume of the collecting chamber will also have an influence on the concentration of chemical in the collecting medium. A larger amount chemical will be permeating through a larger surface area than through a smaller one per unit of time into the same volume (closed loop system) or volume changes per minute (open loop system) of the collecting medium and the BT will decrease (gets shorter) as the MDPR will be reached sooner.[24,29]

In an open system the flow rate will have influence on the concentration and also on the measured BT and PR. At an increasing flow rate the measured BT values will increase (get longer) due to dilution of the chemical in the collecting medium. This will not be the case in a closed system where the MDPR will be reached more rapidly, resulting in a shorter reported BT. In a closed system the possibility of saturation in the collection medium can have influence on the measured SPR value.[29]

E. Evaluation of Test Results

There have been several comparative investigations performed to evaluate the different parameters.

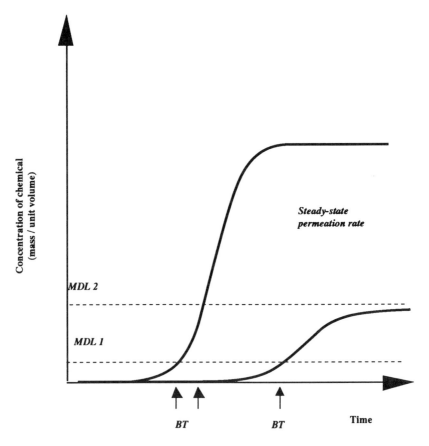

FIGURE 4 Effect of system detection limit in an open loop system. Breakthrough time for materials with different permeabilities.

1. Flow Rate and Mixing

The ASTM F 739-85 version of the standard test method suggested how a homogeneous mixing of the permeated chemical in the collecting medium could be achieved and recommended a flow rate of five collecting chamber volumes through the cell per minute in an open-loop system.[1] The importance of defined flow rate through the collecting compartment was pointed out by the BOHS Technology Committee working party and by Stampfer et al.[10,30] This resulted in an explicit statement for the flow rate of the collecting medium in the ISO/DIS 6529.4 Standard Method. However, the flow rate can be changed to a certain degree before the BT is influenced to a significant degree. Mellström has shown that when using the ASTM cell (1 in.) and the ISO/DIS cell under the same test conditions, the BT was not significantly influenced if the flow of the gas collecting medium varied between 60 to 120 ml/min or between 120 to 500 ml/min, but the PR values was influenced to a significant degree[31,32] (Figure 5).

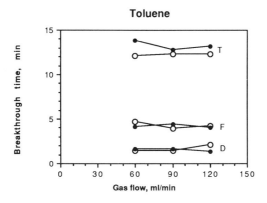

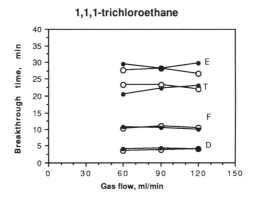

FIGURE 5 Breakthrough times of toluene (top) and 1,1,1-trichloroethane (bottom) through Neoprene® materials at different gas flow rates in the ASTM (white circles) and ISO/DIS (black circles) permeation test cells. D = Dermapren surgical glove, E = Edmont Neoprene® glove, F = Fairprene milled Neoprene® sheet material, T = Neoprene® sheet material made by Trelleborg AB, Sweden. Continuous measurement in an open loop system.

In the current edition of ASTM F 739-91, for an open-loop system there are recommendations of flow rates in the range of 50 to 150 ml/min for the collecting medium. As the volume of the collecting compartment in the ASTM (2 in.) standard test cell is 100 ml, these flow rates are equivalent to 0.5 to 1.5 volume changes per minute, a reduction of the flow rate compared to the recommendations in the earlier edition ASTM F 739-85 (see Table 1).

2. Cell Size and Design

The standard ASTM permeation test cell (2 in.) has been compared to the smaller (1 in.) and found to give equivalent results.[20,33] Henry used air in an open loop system and water in a closed loop system as the collecting media.

BT and PR were compared for both the large and small ASTM permeation test cells. The small test cell gave results not significantly different at the 5% level from the large ASTM test cell.[20] Vahdat used an open loop system with gas flow rates of 1000 ml/min through the large ASTM (2 in.) test cell and 250 ml/min in the small ASTM (1 in.) test cell.[33] Permeation measurements were done with six chemical/glove combinations and the BT and PR values were determined. Statistical analysis of the results showed that the variations of BT and PR between the two test cells were not significant and that the two cells were equivalent.[33] Other permeation test cells of different sizes and design have also been compared with the ASTM standard cell and some of them were considered equivalent. Berardinelli et al. compared the large ASTM test cell and a smaller, commercially available test cell, the AMK cell (AMK Glass Company). This test cell requires less liquid permeant than the ASTM standard test cell and the glove membrane is in a horizontal position. The challenge and collecting compartments both have a volume of 20.4 ml compared to the volumes of the ASTM standard test cell of 45 and 100 ml, respectively. The differences between the BT values measured for the two test cells were not statistically significant, but for the PR values there were significant differences.[18]

3. Open or Closed Loop System

Berardinelli and Moyer have also performed a comparison of BT values obtained with a photoionization detector (PID) in open and closed loop systems, and with an infrared gas analyzer (IR) in a closed system at two different flow rates. They concluded that the measured BT values were system dependent.[15]

4. ASTM vs. ISO/DIS Test Method

As mentioned earlier, there are some main differences between the two permeation test methods, the ASTM F 739-85 and ISO/DIS 6529.4, as can be seen in Table 1. An investigation comparing the chemical permeation data obtained using the ASTM (1 in.) and ISO/DIS cells when performing the testing in accordance with the respective ASTM F 739-85 and ISO/DIS 6529.4 test procedures was done by Mellström.[32] This investigation showed that the measured BT values obtained with the ASTM F 739-85 or ISO/DIS 6529.4 test procedure with the same sampling system are comparable even if the test conditions vary over a wide range as can be seen in Table 3. The PR values and MASS-60 values (mass permeated per square centimeter 60 min after breakthrough) were influenced to a much higher degree by variations in the test conditions and, therefore, more difficult to evaluate. This investigation showed that both BT and PR values differed significantly if different sampling methods (continuous vs. sampling) were used (Figures 6 and 7).

TABLE 3 Breakthrough Time Test Results[a]

Chemical material	Cell	(I) BT min continuous	(II) BT min sampling	(III) BT min sampling
TOL				
Dermapren	ASTM	2.1	2.5	2.6
	ISO	1.3	2.3	2.3
Fairprene	ASTM	4.2	5.6	5.3
	ISO	4.0	5.2	5.2
TCE				
Dermapren	ASTM	4.0	4.7	4.8
	ISO	4.1	4.8	4.8
Fairprene	ASTM	10.6	15.0	14.1
	ISO	10.1	11.5	11.5

Note: Dermapren = surgical Neoprene® glove, 0.23 mm, Fairprene = milled Neoprene® material, 0.46 mm; TOL = toluene, TCE = 1,1,1-trichloroethane.

[a] Results using the ASTM (1 in.) and ISO cells and the (I) ASTM F 739-85 test procedure (F = 120 ml/min, continuous measurement), (II) ISO test procedure (F = 500 ml/min, sampling), (III) ASTM cell in the ASTM F 739-85 Test Procedure and the ISO Cell in the ISO Test Procedure (F = 120 and 500 ml/min sampling, respectively, in open loop systems).

5. Temperature

The test temperature has a crucial influence on the test results and several investigators have shown that the permeation rate increases with the temperature.[17,22,30,34] Vahdat and Bush[34] made chemical permeation measurement for 6 chemical/glove material systems at 5 different temperatures: 25, 35, 45, 55, and 65°C. Their results showed that both BT and PR values vary exponentially with temperature, but BT was more dependent on temperature changes. Information on test temperature should be reported in all standard test methods. This kind of information is of importance when there is intention of handling hot chemicals, handling chemicals in warm or hot environments, and also for firefighters in emergency situations where chemicals are involved.

The manufacturer of the 4H™-glove reported BT results obtained at two different temperatures, room temperature (21°C) and skin temperature (34°C) in their permeation guidelist.[35]

6. Thickness and Formulation of the Material

Note that gloves of the same generic material and the same thickness, but from different manufacturers, can give quite different permeation test results and thus different protection due to the formulation of the dipping mixture, the properties of the polymers, and chemical additives.[36,37] Mickelsen and Hall used nitrile rubber and Neoprene® rubber gloves of similar nominal thickness and each was tested for BT against three chemicals. They found a significant

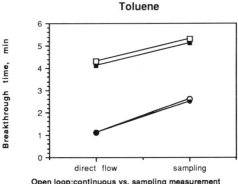

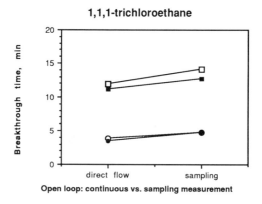

FIGURE 6 Breakthrough times of toluene (top) and 1,1,1-trichloroethane (bottom) through Neoprene® materials. Continuous vs. sampling measurement in an open loop system with gas flow rate of 120 ml/min in the ASTM (white symbols) and ISO/DIS (black symbols) permeation test cells. Circles: Dermapren surgical glove 0.24 mm; squares: Fairprene 0.42 mm (Fairprene Industrial Products, Fairfield, CT), milled Neoprene® sheet material.

difference in chemical BT among generically similar products produced by different manufacturers. The largest difference was obtained for perchloroethylene and nitrile gloves, where the mean BT was 30 vs. 300 min.[36] Sansone et al. tested natural rubber and nitrile gloves against six chemicals; their results also demonstrated that there may be a significant difference in performance between gloves from different manufacturers.[37]

This means that permeation test results such as BT and PR for a certain combination of chemical/glove material cannot be directly translated and used generally for the same chemical and other gloves of the same generic material and nominal thickness, but can be used as guide values.

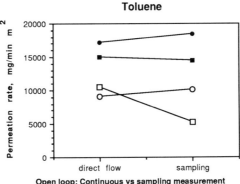

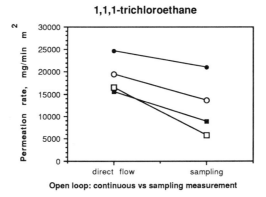

FIGURE 7 Permeation rate of toluene (top) and 1,1,1.trichloroethane (bottom) through Neoprene® materials. Continuous vs. sampling measurement in an open loop system with gas flow rate of 120 ml/min in the ASTM (white symbols) and ISO/DIS (black symbols) permeation test cells. Circles: Dermapren surgical glove 0.24 mm; squares: Fairprene 0.42 mm (Fairprene Industrial Products, Fairfield, CT), milled Neoprene® sheet material.

F. Intermittent Contact Permeation Testing

In actual working situations with hazardous chemicals there is more often a matter of splashes or intermittent contact to deal with rather than continuous contact. The ASTM F 739-85 Standard Test Method involves constant contact, usually during 3 to 4 h, and this could be considered unusually severe for gloves used for protection in cases of accidental splashes or intermittent contact with smaller volumes of chemicals. Man et al. have performed a comparative permeation study of liquid contact, liquid splashes, and vapor influence on the BT. They compared liquid contact for 3 h, intermittent exposure every 15 and 30 min until breakthrough, and a single initial splash. The ASTM procedure was slightly

TABLE 4 Classification of Protective Gloves from Chemical Permeation Test Results (prEN 374-1)

Protection index	Measured breakthrough time
Class 1	>10 min
Class 2	>30 min
Class 3	>60 min
Class 4	>120 min
Class 5	>240 min
Class 6	>480 min

modified and for the intermittent liquid contact the test cell was tilted through the vertical until the test material was horizontal, allowing a small amount (1 to 2 ml) of the chemical to contact and completely cover the material for a period of 2 s. The test cell was then returned to its original position. Two distinct modes of behavior were observed. One, where a more prolonged or concentrated liquid contact resulted in a shorter BT, as expected; in the second mode the BTs of the pure liquid and all splash conditions were essentially the same, and it was the case when the surface was wetted by the liquid and retained in constant contact with the glove material. It was also observed that some of the test chemicals showed one mode of behavior with one material and another mode of behavior with a second material.[38]

This work has been the basis for the development of a modification of permeation testing resulting in the ASTM F 1383 Standard Test Method for Resistance of Protective Clothing Materials to Permeation by Liquids or Gases under Conditions of Intermittent Contact.[14]

G. Protection Index

In the draft standard prEN 374-1, Part 1: Terminology and Performance Requirements, each combination of gloves and test chemical should be classified according to the BT results and given a relative number (see Table 4). This kind of classification can probably facilitate the selection of a protective glove for a certain working situation, but it should be remembered that chemical permeation test data are not the only criteria to be taken into account in the selection process.[39-41]

H. Modified Permeation Testing

The described standard test methods and test cells are useful in most cases of permeation testing, but in some situations modifications are used. For example, when testing highly toxic substances such as pesticides, smaller test cells are preferred to avoid handling large quantities of test chemicals. These can be the same kind of permeation cells that are used for studies of percutaneous absorption *in vitro*.[42] When chemicals with low water solubility or low volatility are to be tested, other collecting media must be used. Solvent mixtures or water with surface active agents added can sometimes be used if proven not to affect the glove material.[43,44] For low volatility pesticide formulations, solid silicon rubber sheets have been used as a collecting medium in a modified ASTM test cell by Ehntholt et al.[45,46] Pinette et al. compared an intermittent collection procedure as an alternative to the solid collection method for nonvolatile water-soluble chemicals. They concluded that the splash collecting procedure had advantages over the solid collecting method as concerning the surface contact and adsorption efficiency. They concluded that a method is needed for testing permeation of water-soluble chemicals having low volatility, but today there is no general procedure available.[47]

Spence describes a trapping system to concentrate low volatility compounds in the sample of the collecting medium in order to avoid the risk of saturation of the gas in the collecting compartment.[48] Lara and Drolet used a modified version of the ASTM F 739 permeation cell in a horizontal position in tests with dynamite gel preparations.[49]

IV. PENETRATION TESTING

Penetration testing is used to determine the flow of liquids through gloves and protective clothing materials on a nonmolecular level. It can be used to investigate the penetration of chemicals and/or microorganisms through porous materials, seams, closures, pinholes, and other imperfections in the glove material. Standard penetration test methods are used as quality control testing for freedom from holes in gloves (leakage test). There is one ASTM standard test method and two prEN standard draft methods for penetration testing of protective gloves and protective clothing materials for resistance to liquid chemicals. There is also one prEN standard draft for a leakage test of medical gloves for single use. These are

- The ASTM F 903-87 Standard Test Method for Resistance of Protective Clothing Materials to Penetration by Liquid Chemicals.[50]
- The prEN 374-2 Protective Gloves Against Chemicals and Microorganisms. Part 2: Determination of Resistance to Penetration. (Two alternative methods are presented, the air leak test and water leak test.)[12]
- The prEN 455-1 Medical Gloves For Single Use. Part 1: Requirement and Testing For Freedom From Holes.[51]

The water leak test in prEN 374-2 is almost identical to that described in prEN-455-1, with only a minor difference between the fill tubs that should be attached to the glove: 1000 ml water shall be added into the glove and then immediately examined for water leaks. If the glove does not leak immediately, a second observation shall be made after another 2 min. If no leak can be observed, the unit passes.

In the air leak test (prEN 374-2) the glove is attached to a circular fixing mandrel and inflated with air under water at room temperature to a given gauge pressure. The mandrel shall be rotated in order to examine the whole glove surface for the emergence of air bubbles.

The two prEN draft standard test methods can be used for standard quality control testing as well as when the whole glove is used, and not only a part. They also have the same requirements concerning the sampling procedure for inspection, with reference to the ISO 2859-1974, Sampling Procedure for Inspections by Attributes.[52] There are three Inspection Levels that determine the relationship between the lot or batch size and the sample size. The Acceptable Quality Level (AQL) is a designated value of defects per hundred units that the consumer indicates will be accepted most of the time by the acceptance sampling procedure to be used. If more than a given number of the tested gloves fail the AQL, the lot or batch will be rejected.

In the ASTM F 903-87 Standard Test Method a circular specimen of the glove is mounted in a penetration test cell, then charged with a challenge liquid under defined pressure. The glove specimen is observed after 5 min in atmospheric pressure and then after 10 min at 13.8 kPa. The appearance of a drop of liquid indicates failure. This method is used to investigate penetration of chemicals through glove materials, closures, and seams connecting gloves to protective suits.[53]

There are several standardized leakage test methods designed for medical gloves, that is, surgical gloves and examination gloves made of rubber or plastic. Six of these, three developed by ASTM and three by the Department of Defense, were evaluated in a laboratory test by Carey et al. and were found in all tests to have inherent limitations.[54] The ability of the water leak test for medical gloves to detect gloves with potential for virus penetration has been studied by Kotilainen et al. Extreme variability in vinyl glove quality, but less in latex gloves, was observed. Gloves that passed a 1000-ml water challenge are unlikely to allow the passage of small viruses.[55-57] Douglas et al. compared standard leak test methods for medical gloves. The testing was done by air inflation, chemical fill, and water fill methods. The methods varied in reproducibility and the chemical fill method detected the highest portion of pinholes.[58]

The standard quality control testing is discussed in detail in Chapter 9 of this volume.

V. SUMMARY

1. The protective effect of gloves against chemicals depends on:
   A. Material quality and formulation
   B. Material thickness

2. The protective effect can be evaluated by:
   A. Degradation testing (screening)
   B. Permeation testing (resistance)
   C. Penetration testing (leakage)
3. A standard test method is a method designed for laboratory testing in a standardized way and is not expected to represent all conditions likely to be found in working situations.
4. The test data should be restricted to comparison of the glove materials mainly on a relative basis.
5. The permeation test results depends to a certain extent of the test conditions such as:
   A. Test cell and design
   B. Collecting medium, volume, and/or flow rate
   C. System mode (open or closed loop)
   D. Continuous or sampling measurement
   E. Sensitivity of the analytical equipment/system
   F. Test temperature
   G. Testing personnel (educated and well trained)
6. The permeation and penetration test data, together with results from mechanical tests and other factors, have to be evaluated in the selection process for the most suitable protective glove for different kinds of working situations.[59]

REFERENCES

1. ASTM F 739-85, Standard Test Method. Resistance of protective clothing materials to permeation by liquids or gases, in *Annual Book of ASTM Standards*, American Society for Testing and Materials, Philadelphia, 1986, 426.
2. European Standard Draft, prEN 374-1. Protective gloves against chemicals and microorganisms. Part 1. Terminology and performance requirements, Final Draft, Comité Européen de Normalisation (CEN/TC 162), April, 1992.
3. European Standard Draft, prEN 369 (E). Protective clothing — protection against liquid chemicals — test method: resistance of materials to permeation by liquids, Final draft, Comité Européen de Normalisation (CEN/TC 162), Jan., 1992.
4. Schwope, A. D., Carroll, T. R., Huang, R., and Royer, M. D., Test kit for field evaluation of the chemical resistance of protective clothing, in *Performance of Protective Clothing: Second Symposium, ASTM STP 989*, Mansdorf, S. Z., Sager, R., and Nielsen, A. P., Eds., American Society for Testing and Materials, Philadelphia, 1988, 314.
5. Stampfer, J. F., Beckman, R. J., and Berardinelli, S. P., Using immersion test data to screen chemical protective clothing, *Am. Ind. Hyg. Assoc. J.*, 49, 579, 1988.
6. Henry, N. W., III, A critical evaluation of protective clothing test methods, in *Proc. 2nd Scand. Symp. Protective Clothing against Chemicals and Other Health Risks, November 1986*, Arbete & Hälsa, 1987, 12, Mellström, G. and Carlsson, B., Eds., National Board of Occupational Safety and Health, Sweden, 1987, 65.

7. Berardinelli, S. P. and Roder, M., Chemical protective clothing field evaluation methods, in *Performance of Protective Clothing: First Symposium, ASTM STP 900,* Barker, R. L. and Coletta, G. C., Eds., American Society for Testing and Materials, Philadelphia, 1986, 250.
8. ASTM F 739-81, Test for resistance of protective clothing materials to permeation by hazardous liquid chemicals, in *Annual Book of ASTM Standards,* American Society for Testing and Materials, Philadelphia, 1981.
9. ASTM F 739-91, Standard Test Method. Resistance of protective clothing materials to permeation by liquids or gases under continuous conditions, in *Annual Book of ASTM Standards,* American Society for Testing and Materials, Philadelphia, April, 1992.
10. BOHS (British Occupational Hygiene Society), Technology committee working party on protective clothing. The development of a standard test method for determining permeation of liquid chemicals through protective clothing materials, *Ann. Occup. Hyg.,* 30, 381, 1986.
11. ISO, Protective clothing — protection against liquid chemicals — determination of resistance of air-impermeable materials to permeation by liquids, Draft International Standard ISO/DIS 6529.4, International Organization for Standardization, TC94/SC13, Genéve, 1, 1988.
12. European Standard Draft, prEN 374-2. Protective gloves against chemicals and microorganisms. Part 2. Determination of resistance to penetration, Comité Européen de Normalisation (CEN/TC 162), Brussels, April, 1992.
13. European Standard Draft, prEN 374-3. Protective gloves against chemicals and microorganisms. Part 3. Determination of resistance to permeation by chemicals, Comité Européen de Normalisation (CEN/TC 162), Brussels, April, 1992.
14. ASTM F 1383, Standard Test Method. Resistance of protective clothing materials to permeation by liquids or gases under conditions of intermittent contact, in *Annual Book of ASTM Standards,* American Society for Testing and Materials, Philadelphia, 1992.
15. Berardinelli, S. P. and Moyer, E. S., Chemical protective clothing breakthrough time. Comparison of several test systems, *Am. Ind. Hyg. Assoc. J.,* 49, 89, 1988.
16. Berardinelli, S. P. and Hall, R., Site-specific whole glove chemical permeation, *Am. Ind. Hyg. Assoc. J.,* 46, 60, 1985.
17. Vahdat, N., Permeation of polymeric materials by toluene, *Am. Ind. Hyg. Assoc. J.,* 48, 155, 987.
18. Berardinelli, S. P., Mickelsen, P., and Roder, M. M., Chemical protective clothing. A comparison of chemical permeation test cells and direct-reading instruments, *Am. Ind. Hyg. Assoc. J.,* 44, 886, 1983.
19. Berardinelli, S. P., Rusczek, R. A., and Mickelsen, R. L., A portable chemical protective clothing test method. Application at a chemical plant, *Am. Ind. Hyg. Assoc. J.,* 48, 804, 1987.
20. Jencen, D. A. and Hardy, J. K., Method for the evaluation of the permeation characteristics of protective glove materials, *Am. Ind. Hyg. Assoc. J.,* 49, 293, 1988.
21. Henry, N. W., III, Comparative evaluation of a smaller version of the ASTM permeation test cell versus the standard cell, in *Performance of Protective Clothing: Second Symposium, ASTM STP 989,* Mansdorf, S. Z., Sager, R., and Nielsen, A. P., Eds., American Society for Testing and Materials, Philadelphia, 1988, 236.

22. Moody, R. P. and Ritter, L., Pesticide glove permeation analysis. Comparison of the ASTM F 739 test method with an automatic flow-through reverse-phase liquid chromatography procedure, *Am. Ind. Hyg. Assoc. J.,* 51, 79, 1990.
23. Schwope, A. D., Costas, P. P., Mond, C. M., Nolen, R. L., Conoley, M., Garcia, D. B., Walters, D. B., and Prokopetz, A. T., Gloves for protection from aqueous formaldehyde: permeation resistance and human factors analysis, *Appl. Am. Ind. Hyg.,* III, 6, 1988.
24. Schwope, A. D., Goydan, R., Reid, R. C., and Krishnamurthy, S., State-of-the-art review of permeation testing and the interpretation of its results, *Am. Ind. Hyg. Assoc. J.,* 49, 557, 1988.
25. Patton, G. L., Conoley, M., and Keith, L.H., Problems in determining permeation cell equivalency, in *Performance of Protective Clothing: Second Symposium, ASTM STP 989,* Mansdorf, S. Z., Sager, R., and Nielsen, A. P., Eds., American Society for Testing and Materials, Philadelphia, 1988, 243.
26. Billing, C. B. and Bentz, A. P., Effect of temperature, material thickness and experimental apparatus on the permeation measurement, in *Performance of Protective Clothing: Second Symposium, ASTM STP 989,* Mansdorf, S. Z., Sager, R., and Nielsen, A. P., Eds., American Society for Testing and Materials, Philadelphia, 1988, 226.
27. Nelson, G. O., Lum, B. B., Carlson, G. J., Wong, C. M., and Johnson, J. S., Glove permeation by organic solvents, *Am. Ind. Hyg. Assoc. J.,* 42, 217, 1981.
28. Jamke, R. A., Understanding and using chemical permeation data in the selection of chemical protective clothing, in *Chemical Protective Clothing Performance in Chemical Emergency Response, ASTM STP 1037,* Perkins, J. L. and Stull, J. O., Eds., American Society for Testing and Materials, Philadelphia, 1989, 11.
29. Kairys, C. J., MDPR — the need for minimum detectable permeation rate requirement in permeation testing of chemical protective clothing, in *Chemical Protective Clothing Performance in Chemical Emergency Response, ASTM STP 1037,* Perkins, J. L. and Stull, J. O., Eds., American Society for Testing and Materials, Philadelphia, 1989, 265.
30. Stampfer, J. F., McLeod, M. J., Martinez, A. M., and Berardinelli, S. P., Permeation of eleven protective garment materials by four organic solvents, *Am. Ind. Hyg. Assoc. J.,* 45, 642, 1984.
31. Mellström, G. A., Comparison of chemical permeation data obtained with ASTM and ISO permeation test cells. I. The ASTM standard test procedure, *Ann. Occup. Hyg.,* 35, 153, 1991.
32. Mellström, G. A., Comparison of chemical permeation data obtained with ASTM and ISO permeation test cells. II. The ISO standard test procedure, *Ann. Occup. Hyg.,* 35, 167, 1991.
33. Vahdat, N., Permeation of polymeric materials by chemicals. A comparison of the 25-mm and 51-mm ASTM-cells, in *Performance of Protective Clothing: Second Symposium, ASTM STP 989,* Mansdorf, S. Z., Sager, R., and Nielsen, A. P., Eds., American Society for Testing and Materials, Philadelphia, 1988, 219.
34. Vahdat, N. and Bush, M. M., Influence of temperature on the permeation properties of protective clothing material, in *Chemical Protective Clothing Performance in Chemical Emergency Response, ASTM STP 1037,* Perkins, J. L. and Stull, J. O., Eds., American Society for Testing and Materials, Philadelphia, 1989, 132.

35. Chemical Protection List, 4H™-Glove, Safety 4 A/S Lyngby, Denmark, June, 1989.
36. Mickelsen, R. L. and Hall, R. C., A breakthrough time comparison of nitrile and neoprene glove materials produced by different glove manufacturers, *Am. Ind. Hyg. Assoc. J.,* 48, 941, 1987.
37. Sansone, E. B. and Tewari, Y. B., Differences in the extent of solvent penetration through natural rubber and nitrile gloves from various manufacturers, *Am. Ind. Hyg. Assoc. J.,* 41, 527, 1980.
38. Man, V. L., Bastecki, V., Vandal, G., and Bentz, A., Permeation of protective clothing materials. Comparison of liquid contact, liquid splashes, and vapors on breakthrough times, *Am. Ind. Hyg. Assoc. J.,* 48, 551, 1987.
39. Perkins, J. L., Chemical protective clothing. I. Selection and use, *Appl. Ind. Hyg.,* 2, 222, 1988.
40. Mellström, G. A., Protective gloves of polymeric materials. Experimental permeation testing and clinical study of side effects, *Arbete och Hälsa,* National Board of Occupational Safety and Health, Stockholm, 15, 1991.
41. Leinster, P., Bonsall, J. L., Evans, M. J., and Lewis, S. J., The application of test data in the selection and use of the gloves against chemicals, *Ann. Occup. Hyg.,* 34, 85, 1990.
42. Colligan, S. A. and Horstman, S. W., Permeation of cancer chemotherapeutic drug through glove materials under static and flexed conditions, *Appl. Occup. Environ. Hyg.,* 5(12), 848, 1990.
43. Que Hee, S. S., Permeation of some pesticidal formulations through glove materials, in *Chemical Protective Clothing Performance in Chemical Emergency Response, ASTM STP 1037,* Perkins, J. L. and Stull, J. O., Eds., American Society for Testing and Materials, Philadelphia, 1989, 157.
44. Ehntholt, D. J., Cerundolo, D. L., Bodek, I., Schwope, A. D., Royer, M. D., and Nielsen, A. P., A test method for the evaluation of protective glove materials used in agricultural pesticide operations, *Am. Ind. Hyg. Assoc. J.,* 51, 462, 1990.
45. Ehntholt, D. J., Almeida, R. F., Beltis, K. J., Cerundolo, D. L., Schwope, A. D., Whelan, R. H., Royer, M. D., and Nielsen A. P., Test method development and evaluation of protective clothing items used in agricultural pesticide operations, in *Performance of Protective Clothing: Second Symposium, ASTM STP 989,* Mansdorf, S. Z., Sager, R., and Nielsen, A. P., Eds., American Society for Testing and Materials, Philadelphia, 1988, 727.
46. Ehntholt, D. J., Bodek, I., Valentine, J. R., Schwope, A.D., Royer, M.D., Frank, U., and Nielsen, A. P., The effects of solvent type and concentration on the permeation of pesticide formulations through chemical materials protective glove, in *Chemical Protective Clothing Performance in Chemical Emergency Response, ASTM STP 1037,* Perkins, J. L. and Stull, J. O., Eds., American Society for Testing and Materials, Philadelphia, 1989, 146.
47. Pinette, M. F. S., Stull, J. O., Dodgen, C. R., and Morley, M. G., A preliminary study of an intermittent collection procedure as an alternative permeation method for nonvolatile, water soluble chemicals, in *Performance of Protective Clothing: Fourth Volume, ASTM STP 1133,* McBriarty, J. P. and Henry, W. N., Eds., American Society for Testing and Materials, Philadelphia, 1992, 339.

48. Spence, M. W., An analytical technique for permeation testing of compounds with low volatility and water solubility, in *Performance of Protective Clothing: Second Symposium, ASTM STP 989,* Mansdorf, S. Z., Sager, R., and Nielsen, A. P., Eds., American Society for Testing and Materials, Philadelphia, 1988, 277.
49. Lara, J. and Drolet, D., Testing the resistance of protective clothing materials to nitroglycerin and ethylene glycol dinitrate, in *Performance of Protective Clothing: Fourth Volume, ASTM STP 1133,* McBriarty, J. P. and Henry, W. N., III, Eds., American Society for Testing and Materials, Philadelphia, 1992, 153.
50. ASTM F 903-87, Standard Test Method. Resistance of protective clothing materials to penetration by liquids, in *Annual Book of ASTM Standards,* American Society for Testing and Materials, Philadelphia, 1988.
51. European Standard Draft, prEN 455-1: Medical gloves for single use. Part 1. Requirement and testing for freedom from holes, Comité Européen de Normalisation (CEN/TC 205/WG 3), July, 1991.
52. ISO, International Standard ISO 2859-1974 (E): Sampling procedures and tables for inspection by attributes, International Organization for Standardization, 1974.
53. Berardinelli, S. P. and Cottingham, L., Evaluation of chemical protective garment seams and closures for resistance to liquid penetration, in *Performance of Protective Clothing: First Symposium, ASTM STP 900,* Barker, R. L. and Coletta, G. C., Eds., American Society for Testing and Materials, Philadelphia, 1986, 263.
54. Carey, R., Herman, W., Herman, B., and Casamento, J., A laboratory evaluation of standard leakage tests for surgical and examination gloves, *J. Clin. Eng.,* 14, 133, 1989.
55. Kotilainen, H. R., Brinker, J. P., Avato, J. L., and Gantz, N. M., Latex and vinyl examination gloves. Quality control procedures and implications for health care workers, *Arch. Intern. Med.,* 149, 2749, 1989.
56. Kotilainen, H. R., Avato, J. L., and Gantz, N. M., Latex and vinyl nonsterile examination gloves: status report on laboratory evaluation of defects by physical and biological methods, *Appl. Environ. Microbiol.,* June, 1627, 1990.
57. Kotilainen, H. R., Cyr, W. H., Truscott, W., Nelson, M. G., Routson, L. B., and Lytle, C. D., Ability of 1000 mL water leak test for medical gloves to detect gloves with potential for virus penetration, in *Performance of Protective Clothing: Fourth Volume, ASTM STP 1133,* McBriarty, J. P. and Henry, W. N., Eds., American Society for Testing and Materials, Philadelphia, 1992, 38.
58. Douglas, A. A., Neufeld, P. D., and Wong, R. K. W., An interlaboratory comparison of standard test methods for medical gloves, in *Performance of Protective Clothing: Fourth Volume, ASTM STP 1133,* McBriarty, J. P. and Henry, W. N., Eds., American Society for Testing and Materials, Philadelphia, 1992, 99.
59. Stull, J. O., White, D. F., and Greimel, T. C., A comparison of the liquid penetration test with other chemical resistance tests and its application in determining the performance of protective clothing, *Performance of Protective Clothing: Fourth Volume, ASTM STP 1133,* McBriarty, J. P. and Henry, W. N., Eds., American Society for Testing and Materials, Philadelphia, 1992, 123.

7

Chemical Resistance Field Test Methods

Stephen P. Berardinelli, Sr.

TABLE OF CONTENTS

I. Introduction 79
II. Field Test Methods 80
III. Degradation Test Methods 80
IV. Penetration Test Methods 82
V. Permeation Test Methods 82
    A. Liquid Chemical With a Vapor Pressure of 10 torr (10 mmHg) or Above 83
    B. Liquid Chemical With a Vapor Pressure Less Than 10 torr (10 mmHg) 84
    C. Gas or Vapor 84
    D. Detector Tube Method 86
    E. Permeation Cup Test Method 86
VI. Whole-Garment Field Test Methods 87
    A. Whole-Glove Field Test Method 87
    B. Standard Practice for Pressure Testing of Gas-Tight Totally Encapsulating Chemical-Protective Suits (TECP) 87
VII. Conclusions 88
References 89

I. INTRODUCTION

Chemical protective clothing (CPC) is comprised of garments, gloves, boots, coveralls, aprons, or fully encapsulating suits that are designed to protect workers from dermal exposure to chemicals. Even when CPC is worn, exposure can occur as a result of: (1) material failure resulting from chemical or physical degradation, or a combination of both; (2) bulk penetration of liquid chemical through pinholes, seams, closures, imperfections, and other discontinuities; and (3) permeation or molecular flow of liquid chemical through the garment material. ASTM Committee F-23 on Protective Clothing, which develops consensus standard methods, has promulgated standard test methods for evaluating chemical permeation and penetration: *ASTM Test for Resistance of Protective Clothing Materials to Permeation by Liquid Chemicals* (F 739), and *ASTM Test for Resistance of Protective Clothing to Penetration by Liquid Chemicals* (F 903). These standard methods are designed for the laboratory rather than for field use. This need for laboratory tests can delay the acquisition of important information needed by health and safety professionals for the proper selection and use of CPC.

II. FIELD TEST METHODS

It is best that CPC be tested under anticipated or actual work conditions. This testing is necessary because of the differences in products and barrier materials and because of the inability to predict mixture permeation characteristics from pure chemical data.[1,2]

Reuse of any CPC after decontamination is not recommended unless chemical resistance testing is conducted after decontamination. Testing should evaluate: (1) any adverse effect of the decontamination procedure on the CPC material, and (2) the effectiveness of the decontamination procedure in removing the chemical from the material.[3,4] Chemical and physical resistance tests should also be repeated after the CPC has undergone a number of chemical exposure and decontamination cycles. Field evaluations could include the following:

1. Degradation, penetration, or permeation testing, or a combination of these tests. Tests should utilize the actual chemicals to be encountered on specimens from the products being considered, under the expected conditions (contact sequence, temperature, and reuse).
2. Simple observations of characteristics of use in the field. Important use factors include: Does the CPC introduce other hazards such as loss of dexterity or snagging on moving equipment? Do workers accept and use the CPC? Does additional exposure occur from donning, recycling, or storing the CPC? Do exposure indications (skin disorders, symptoms, positive blood tests) increase or decrease when workers wear CPC? What are the economic incentives vs. potential exposure risks with reuse of the garments (decontamination and reuse vs. single use and disposal)?
3. Worker information on CPC use and acceptance (fit, comfort, etc.).

III. DEGRADATION TEST METHODS

The resistance of CPC to chemical degradation can be evaluated by an easily performed, inexpensive test method.[5] This test method determines the resistance of protective clothing materials to degradation by liquid chemicals under

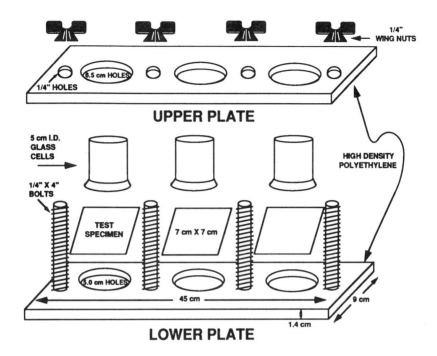

FIGURE 1 Chemical degradation test apparatus.

the condition of continuous liquid contact. Essentially, it is a measure of chemical compatibility between a fabric and a challenge liquid.

In this test, one side of a CPC material specimen is exposed to a challenge liquid for a known period of time. At the end of this known time, changes in visual appearance, thickness, and weight of the clothing material specimen are observed and measured.

A test apparatus (Figure 1) and procedure to determine weight and thickness have been described elsewhere.[6] Visual changes in appearance are described in subjective terms (e.g., bleached, swollen, disintegrated). The performance of each specimen is determined using the following scale:

Rating	Visual changes in sample after
1	5 min
2	30 min
3	1 h
4	4 h

Chemical protective materials that receive a scale rating of 1 or 2, or exhibit more than 20% weight or thickness change, are considered unacceptable; no further testing for penetration or permeation is warranted. Chemical degradation data do not indicate chemical penetration or permeation. However, degradation testing can be useful in the elimination of unsuitable candidate materials for penetration and permeation testing.

IV. PENETRATION TEST METHODS

One method for testing CPC resistance to chemical penetration is a modified version of the ASTM F-23 Committee F 903 Standard Test Method.[6] Test specimens without discontinuities, such as zippers and seams, should be evaluated for quality assurance (e.g., pinholes). Specimens with zippers and seams should be tested as well, to evaluate their penetration resistance.

The resistance of a protective clothing material to penetration by a liquid is determined visually; that is, appearance of liquid droplets through a material at atmospheric pressure (after 5 min) and then at 6.9 kPa pressure (1 psig for 10 min). The test apparatus, shown in Figure 2, has been described previously.[7] A "lecture bottle" of compressed air is sufficient to do 30 tests. An iterative procedure for conducting this test has also been described previously.[8]

Penetration test results should be reported as pass or fail. Specimens leaking within the 5-min period at atmospheric pressure may be reported as failing during atmospheric testing. Specimens that leak during the pressure test may be reported as failing during pressure testing. The specimen must pass both the atmospheric and pressure test to be acceptable.

V. PERMEATION TEST METHODS

Chemical permeation field tests are the most complicated field tests. In the permeation test, the clothing material acts as a barrier separating the challenge chemical from a collection medium which captures the permeating chemical. The collection medium, usually air, is analyzed to measure the concentration of challenge chemical and, thereby, the amount of hazardous chemical that has permeated the barrier as a function of time.

Specifically, the resistance of a protective clothing material to permeation by a liquid chemical is characterized by measuring two parameters: the breakthrough time and the subsequent steady-state rate at which the liquid permeates the clothing material.

Breakthrough time (BT) may be used to estimate how long the CPC provides protection while in continuous contact with the test liquid. BT is defined as the elapsed time between initial contact of the hazardous liquid chemical with the outside surface of a protective clothing material and the time at which the chemical can be detected at the inside surface of the material by means of the analytical technique.

Steady-state permeation is the constant rate of permeation that occurs after the breakthrough, when all forces affecting permeation have reached equilibrium. It does not occur for some chemical-material testing combinations. If the

Chemical Resistance Field Test Methods

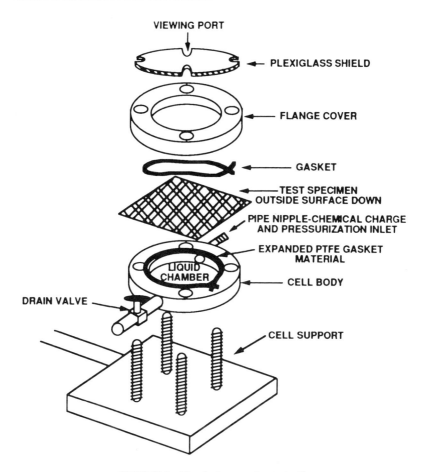

FIGURE 2 Chemical penetration test cell.

BTs of two products are similar, the one with the lower steady-state permeation rate would allow less chemical exposure if the product were to be used beyond its BT, and hence would be the best selection.

Selecting a chemical permeation field test method depends on the challenge chemical state (gas, liquid, solid) and the vapor pressure. The methods described here are good examples; other methods may perform equally well.

A. Liquid Chemical With a Vapor Pressure 10 torr (10 mmHg) or Above

The arrangement of this field test system is shown as Figure 3 and discussed in Reference 9. The test system consists of a two-chambered permeation cell (Figure 4) which holds the test specimen, a direct-reading instrument which detects chemicals, and a portable pump to circulate the collection medium. Many different direct-reading instruments can be used. The following instruments have been used with good results:

1. *Photoionization detector*. This instrument is nonspecific for analysis; that is, it does not separate chemicals. The observed detector response could be caused by the presence of a single gaseous chemical or several gaseous chemicals. However, it does indicate breakthrough.
2. *Organic vapor analyzer*. This instrument uses a flame ionization detector that functions as a total organic vapor analyzer.

Typically, a test is run for 4 h, which is a sufficient period to graph time against concentration (detector meter reading). Longer or shorter test times may be selected based on the potential exposure time. A recorder connected to the direct-reading detector can be used to plot meter readings vs. time quite conveniently.

B. Liquid Chemical With a Vapor Pressure Less Than 10 torr (10 mmHg)

In this test, no cell is used. A finger or swatch of material formed into a pouch is filled with liquid, then sealed. An electrical cable tie works well. The outside of the material is swiped at specified times with a small piece of wipe moistened in a solvent that does not degrade the test material. The wipe is extracted then analyzed. Gas chromatography is typically used as the analyzer.[10]

C. Gas or Vapor

The arrangement of this field test system is shown in Figure 5 and discussed in Reference 11. A direct reading instrument, such as a Miran 1A General Purpose Infrared Gas Analyzer (IR), is used as the analyzer. The Miran is a single-beam, variable filter spectrometer capable of scanning the infrared spectral range between 2.5 and 14.5 µm and is equipped with a gas cell having

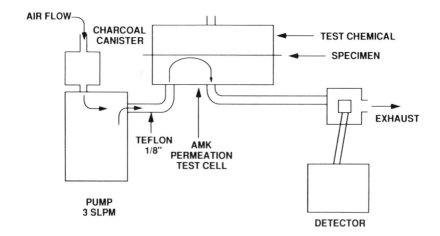

FIGURE 3 Liquid chemical permeation test system.

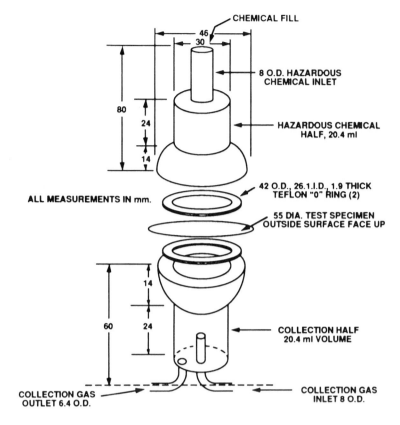

FIGURE 4 Chemical permeation test cell.

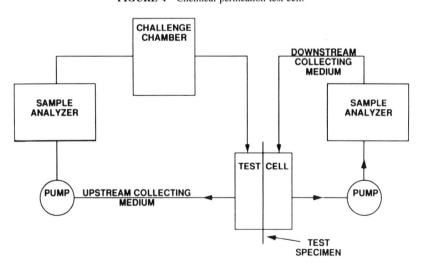

FIGURE 5 Vapor permeation test system.

a variable path length between 0.75 and 20.25 m. As shown in Figure 5, two analyzers are needed — one for the upstream or vapor gas challenge loop, the other for the downstream or permeant loop. The test cell in Figure 4 is used. Two stainless steel pumps, producing a flow rate of approximately 8 l/min, are used to circulate the gas in the two loops. As shown in Figure 5, the upstream loop is purged with dry air and the contaminant introduced by means of a syringe. The upstream loop is then allowed to equilibrate. The upstream vapor concentration is continually monitored by means of the IR detector. Additional contaminant is introduced into the upstream loop until the desired concentration is obtained. Immediately downstream of the specimen is another closed loop that contains the other IR detector to monitor the breakthrough permeation concentration as a function of exposure time.

The specimen is placed in the permeation cell, the cell is connected to the downstream closed loop, and the timer is actuated. Downstream data (concentration of permeant vs. time) are plotted. The BT is determined from the graph.

D. Detector Tube Method

This field test provides a measure of the breakthrough time using a relatively inexpensive colorimetric gas detector tube. The chemical detected must be volatile and must provide a visible indication on the detector tube. Also, the detector tube must be sensitive enough to detect the chemical at breakthrough.[12,13] This test system is depicted in Figure 3, assuming a detector tube is used in lieu of a direct-reading instrument. The Draeger Polytest tube, model CH 28401, was evaluated and appears to be sensitive enough to detect many hydrocarbons. It is based on a generalized iodine peroxide reaction. Other gas detector tube manufacturers provide similar tubes.

For this test, an unused tube is used for comparison, since a faint stain may be detected more easily. Because of the sensitivity of the Draeger Polytest tube, a narrow band of light gray, green, or tan may develop very early in the test. If it does not increase in size or change color, this is probably contamination and should not be considered breakthrough. Contamination could be in the air source or outgassing from the material sample. At breakthrough, a stain will usually deepen to a dark brown, green, or purple color and rapidly begin to lengthen.[12]

E. Permeation Cup Test Method

Another permeation test method applicable to field use employs weight loss of volatile chemical permeants.[14,15] The resistance of a protective clothing material to permeation by a test chemical is assessed by measuring BT and permeation rate through replicate specimens of the material. A clothing material specimen is secured over the mouth of a special shallow cup that holds the test chemical. The outside surface of the material faces the chemical; the inside is open to the atmosphere. Detection of permeation requires that the test chemical have sufficient volatility, approximately 100 torr (100 mmHg).

For this test, the cup assembly is weighed, inverted, and reweighed at predetermined time intervals to determine the amount of chemical that permeates the material and subsequently evaporates to the atmosphere. An analytical balance, readable and reproducible to at least ±1 mg is required for measuring the weight loss of the permeation cup. The capacity of the balance should be at least 50 g more than the weight of an empty permeation cup. The sensitivity of the chemical permeation method increases if the sensitivity of the balance is increased; also, if the time between weighings and the length of the test are increased.

VI. WHOLE-GARMENT FIELD TEST METHODS

A. Whole-Glove Field Test Method

Most chemical resistance standard methods evaluate only a part of the whole protective clothing. To test glove materials, for example, a specimen is usually cut from the palm, back, or gauntlet. In the past, whole-glove chemical resistance testing was performed by measuring a physical parameter such as glove weight, after which the glove was filled with the challenge liquid. After a specified time, the glove was drained of the liquid and patted dry, and the physical parameters were remeasured.

One whole-glove test method uses an inverted glove and a portable direct-reading instrument. A photoionization detector with a recorder may be used to measure the permeant concentration. A glove is turned inside out, approximately 100 ml of challenge liquid is added, and the neck of the glove is tied off with an electrical cable tie. The tied glove neck is evaluated for a tight seal with the photoionization detector probe; then the probe is moved to monitor glove sites. The sites are noted on the recorder chart.[16]

BT is calculated from the recording's trace, as the recorder chart speed is a constant. The mean challenge concentration under steady-state equilibrium at all specific sites is calculated by averaging four concentrations after steady-state equilibrium is achieved.[16]

This method can be used in the field, provided the chemical has sufficient vapor pressure and gives a signal on the photoionization detector. This is easily determined by passing the probe tip over the liquid. Should it not respond, another quick-response analytical instrument could be substituted.

B. Standard Practice For Pressure Testing of Gas-Tight Totally Encapsulating Chemical-Protective Suits (TECP)

This ASTM practice measures the integrity of the suit and visor material, seams, and suit closures of a TECP suit with a positive pressure test.[17] Suit-venting valves must be sealed for this test, and therefore are not functionally tested.

The TECP suit is visually inspected and modified for the test. The test apparatus is attached to the suit to permit inflation to a pretest suit expansion pressure as recommended by the manufacturer for removal of suit wrinkles and creases. The pressure is lowered to the manufacturer's recommended test pressure and monitored for 3 min. At the end of 3 min the pressure is determined again. If the pressure drops more than 20%, the TECP suit fails the test and is removed from service.

If the suit fails the test, the suit should be checked for leaks by inflating to the pretest expansion pressure and brushing or wiping the entire suit (including seams, closures, lens gaskets, glove-to-sleeve joints, etc.) with a mild soap and water solution. The suit should then be observed for the formation of soap bubbles, which is an indication of a leak.

VII. CONCLUSIONS

Quick on-site protective clothing test methods were described that evaluate chemical resistance of protective clothing for degradation, penetration, and permeation. A pressure test for whole garments was also described. Valuable and timely chemical protective clothing information may be obtained by health and safety professionals with the capability to conduct chemical resistance testing on-site.*

REFERENCES

1. Mickelsen, R.L. and Hall, R.C., A breakthrough time of nitrile and neoprene glove materials produced by different glove manufacturers, *Am. Ind. Hyg. Assoc. J.,* 48, 941, 1987.
2. Mickelsen, R.L., Roder, M.M., and Berardinelli, S.P., Permeation of chemical protective clothing by three binary solvent mixtures, *Am. Ind. Hyg. Assoc. J.,* 47, 236, 1986.
3. Vahdat, N. and Delaney, R., Decontamination of chemical protective clothing, *Am. Ind. Hyg. Assoc. J.,* 50, 152, 1989.
4. Perkins, J.L., Decontamination of protective clothing, *Appl. Occup. Environ. Hyg.,* 6, 29, 1991.
5. Coletta, G.C., Mansdorf, S.Z., and Berardinelli, S.P., Chemical protective clothing standard test method development. Part II. Degradation test methods, *Am. Ind. Hyg. Assoc. J.,* 49, 26, 1988.
6. Berardinelli, S.P., Test method for resistance of protective clothing material to degradation by liquid chemicals, in *A Guide for Evaluating the Performance of Chemical Protective Clothing,* Roder, M.M., Ed., U.S. Department of Health and Human Services, National Institute for Occupational Safety and Health, Cincinnati, OH, 1990, C-12.

---

* Mention of brand names does not constitute endorsement by the National Institute for Occupational Safety and Health, the Centers for Disease Control, the U. S. Public Health Service, or the Department of Health and Human Services.

7. Mansdorf, S.Z. and Berardinelli, S.P., Chemical protective clothing standard test method development. Part I. Penetration test method, *Am. Ind. Hyg. Assoc. J.,* 49, 21, 1988.
8. Berardinelli, S.P., Test method for resistance of protective clothing to penetration by liquid chemicals, in *A Guide for Evaluating the Performance of Chemical Protective Clothing,* Roder, M.M., Ed., U.S. Department of Health and Human Services, National Institute for Occupational Safety and Health, Cincinnati, OH, 1990, C-16.
9. Berardinelli, S.P., Rusczek, R.A., and Mickelsen, R.L., A portable chemical protective clothing test method. Application at a chemical plant, *Am. Ind. Hyg. Assoc. J.,* 48, 804, 1987.
10. Stampfer, J.F., McLeod, M.J., Betts, M.R., Martinez, A.M., and Berardinelli, S.P., Permeation of polychlorinated biphenyls and solutions of these substances through selected protective clothing materials, *Am. Ind. Hyg. Assoc. J.,* 45, 634, 1984.
11. Berardinelli, S.P., Test method for resistance of protective clothing materials to permeation by gas or vapor, in *A Guide for Evaluating the Performance of Chemical Protective Clothing,* Roder, M.M., Ed., U.S. Department of Health and Human Services, Cincinnati, OH, 1990, C-29.
12. Roder, M.M. and Hall, R.C., A simplified test method for measuring breakthrough of volatile chemicals through chemical protective clothing, in *A Guide for Evaluating the Performance of Chemical Protective Clothing,* Roder, M.M., Ed., U.S. Department of Health and Human Services, National Institute for Occupational Safety and Health, Cincinnati, OH, 1990, C-1.
13. Sarner, S.F. and Henry, N.W., III, The use of detector tubes following ASTM method F739-85 for measuring permeation resistance of clothing, *Am. Ind. Hyg. Assoc. J.,* 50, 298, 1989.
14. Mickelsen, R.L., Hall, R.C., Chern, R.T., and Myers, J.R., Evaluation of a simple weight-loss method for determining the permeation of organic liquids through rubber films, *Am. Ind. Hyg. Assoc. J.,* 52, 445, 1991.
15. Permeation Cup Test Kits, Kits are available from Arthur D. Little, Inc., 15 Acorn Park, Cambridge, MA, 02140; and TRI, Inc., 9063 Bee Caves Rd., Austin, TX, 78733.
16. Berardinelli, S.P. and Hall, R.C., Site-specific whole glove chemical permeation, *Am. Ind. Hyg. Assoc. J.,* 46, 60, 1985.
17. ASTM Standard Method, F1052-87, *Standard Practice for Pressure Testing of Gas-Tight Totally Encapsulating Chemical-Protective Suits,* ASTM Standards on Protective Clothing, American Society for Testing and Materials, Philadelphia, PA, 1990, 235.

8

Percutaneous Absorption Studies in Animals

Anders S. Boman and Gunh A. Mellström

TABLE OF CONTENTS

I. Introduction 91
II. Study Design 92
    A. *In Vitro* Permeation Testing 93
    B. *In Vivo* Testing 94
III. Results 95
    A. *In Vitro* Permeation Testing 95
    B. *In Vivo* Testing 98
IV. Discussion 100
V. Conclusion 106
References 106

I. INTRODUCTION

Chemicals and products handled in industry include a variety of organic solvents and solvent-containing products such as degreasing fluids, cleansing agents, paints, and plastics. After topical application, all solvents can give rise to side effects on and in the skin, such as defatting, erythema, edema, and irritant contact dermatitis.[1-4] The majority of solvents have low acute systemic toxicity, but skin absorption may add to the body burden absorbed via the respiratory pathway. A few extremely toxic agents are readily absorbed through the skin and therefore need extreme precautions when handled.[5]

    To reduce skin exposure to topically and systemically harmful chemicals, a number of synthetic protective barriers have been developed. These include polymeric membranes in gloves and other protective clothing. The choice of gloves may be random or may be preceded by an assessment of available protective materials based on data from technical testing of polymeric membranes[6,7] or on *in vivo* evaluation of their protective effect.

Chemical permeation testing of protective gloves and clothing is generally performed in a technical system according to accepted or proposed standards, for example the ASTM F 739-85 Standard Test Method or the draft International Standard ISO/DIS 6529.4.[8,9] The parameters of interest in *in vitro* permeation testing are breakthrough time (BT) and steady-state permeation rate (SPR) of the chemicals, and the results from *in vitro* tests are often the only basis for the selection of adequate protective equipment. Little concern is given to the biological implications of these results. However, these technical systems do not consider the effects of the protective materials on the target they are designed to protect, i.e., the occlusive influence they have on the skin. Occlusion is considered to be one of the most crucial factors in increasing percutaneous absorption.[10]

Additional information on protective efficacy can be derived from *in vivo* testing in man or in experimental animals.[11-20] In the literature only a few reports on *in vivo* testing of the efficacy of gloves can be found. These investigations are often carried out in humans, as laboratory investigations,[12] as field studies in the work environment,[13,14,20] or as clinical studies.[15-19] Although the most accurate and reliable data are derived from experiments in humans, there is a certain risk of side effects among the subjects.[12,13]

For screening purposes and to reduce the need for human experiments, animal models may be used for the comparative investigation of the protective effects of gloves.[11] As *in vivo* testing is both expensive and time consuming, it is essential to know if there is a definite relationship between *in vivo* and *in vitro* test results. It would also be of practical use for this knowledge to be available when the most suitable gloves for certain working situations are to be selected and evaluated. Today most of the data available on the protective effect of gloves are obtained *in vitro*.[21,22] In industry, deliberate exposure of the skin to known chemicals is often of rather short duration, but repeated at varying intervals. This is in contrast to accidental and catastrophic pollution where the exposure can be massive.

The aim of this chapter is to describe a method for the biological assessment of the efficacy of protective gloves against percutaneous absorption of organic solvents in the guinea pig. A second aim is to compare the results with protective glove materials obtained through *in vivo* testing in experimental animals with *in vitro* studies using a proposed standard procedure[9] where BT and permeation/absorption rate were the parameters determined.

II. STUDY DESIGN

In the *in vitro* testing, the investigated glove materials were butyl rubber, (Norton butyl, Allt i Skydd, Malmö, Sweden); PVC (Vinylprodukter, Landskrona, Sweden and KEBO Care, Spånga, Sweden), and natural rubber latex (LIC-Hygien, Solna, Sweden). Pieces were cut from the gloves and three thickness measurements (±0.01 mm) were done using an Oditest micrometer

TABLE 1 Glove Materials Tested

Material	Type	n	Thickness, mm (mean ± SD)
PVC (vinyl)	Household glove	12	0.520 ± 0.062
Butyl rubber	Industrial glove	12	0.412 ± 0.014
Natural rubber (latex)	Disposable glove	12	0.195 ± 0.012
PVC (vinyl)	Disposable glove	12	0.143 ± 0.010

From Mellström, G. and Boman, A., *Contact Derm.*, 26, 120, 1992. With permission.

(H.C. Kröplin, GmbH, Schleutern, Germany). Before being mounted into the permeation cell, the specimens were inspected for pinholes or other defects. At least three runs — as stated in the proposed standard method — were performed for each combination of glove and solvent. The testing was carried out at room temperature, $21 \pm 2°C$. Tested glove materials are summarized in Table 1. The solvents used for exposure were *n*-butanol, toluene (Merck, Darmstadt, Germany), and 1,1,1-trichloroethane (Fluka, Buchs, Switzerland), all of *pro analysi* grade.

A. *In Vitro* Permeation Testing

Permeation testing was carried out according to the proposed ISO standard with the ISO permeation test cell and using air as the collecting gas at a flow rate of 500 ml/min in an open-loop system.[9] The air was sampled using an automatic gas sampling valve, and injected into a gas chromatograph (Shimadzu 9A) equipped with a flame ionization detector. N_2 was used as the carrier gas and 20% SE-30 on Chromosorb WHP as the stationary phase. Samples were taken following a time schedule, every 2 min for 15 min and then every 5 min. The test data were recorded and calculated using a Cromatopac data processor (Shimadzu CR3A) connected to the system. The minimum detectable level (MDL) in the *in vitro* study was for butanol — 19.7 $\mu g/cm^2$, for toluene — 0.13 $\mu g/cm^2$, and for trichloroethane — 0.39 $\mu g/cm^2$.

The BT was recorded as the time (in minutes) required for the solvents to appear on the opposite side of the membrane in a concentration exceeding the minimum detection level.[22]

The SPR was reached when the permeant concentration was constant vs. time; usually within 60 to 120 min after breakthrough. It was calculated according to the ISO/DIS procedure, where the cumulative amount (in micrograms per square centimeter) is plotted vs. time and the inclination of the curve gives the permeation rate at steady state.[9]

The permeation rates of the solvents through the glove membranes were related to the rate through natural latex rubber, which was set to 1.00 to give a relative permeation rate (RPR).

B. *In Vivo* Testing

The same glove materials (Table 1) as in the *in vitro* test were investigated in the *in vivo* system, except for the natural rubber glove which was of the same type and thickness except from another manufacturer.[11]

The testing was carried out in guinea pigs according to a method in which the skin of the backs of guinea pigs was exposed with or without protective glove specimens.[23] Female albino or pigmented guinea pigs weighing 500 to 700 g were used. The hair on the back was carefully removed with electric clippers, care being taken not to injure the skin. The animals were then anesthetized by i.p. administration of 36 mg/kg body weight sodium pentobarbital (Mebumal vet., ACO, Solna, Sweden). A polyethylene catheter was inserted into the carotid artery after it had been exposed. Blood for preexposure blank value and individual reference was withdrawn. Heparin was administered to the animals to avoid coagulation.

For the testing with butanol and gloves, a group where the skin had been stripped with cellotape and exposed was used.[11] The glove membranes were glued to the underside of a glass ring, inner diameter 20 mm, with a cyanoacrylate glue (Cyanolit 201, E.A.N. Bennetter, Stockholm, Sweden) and tested for leakage. The glass ring was then glued to the skin. Neat organic solvent (1.0 ml) was placed in the depot so formed, and this was then sealed with a cover glass. Exposure was continuous for 4 to 6 h. The percutaneous absorption rate was based on determination of the solvent blood concentration. Blood samples (0.5 ml) were withdrawn via the catheter inserted in the carotid artery every 10 min during the first 100 min, then at 20- and 30-min intervals. The blood loss was compensated for with Macrodex® (Pharmacia AB, Uppsala, Sweden). The blood concentration of solvent was analyzed with a gas chromatograph using a head space technique, with nitrogen as the carrier gas, methyl silicone SE-30 on Chromosorb as the stationary phase, and a flame ionization detector. The analytical conditions for the solvents in the *in vivo* exposure are presented in Table 2.

Animals exposed to solvent without glove membranes served as controls. A second control group was added for the experiments with butanol, where the skin had been stripped with cellotape until a wet, glistening surface was obtained.

The blood concentration figures from the *in vivo* study were recalculated to give RAR according to a method recently described by Boman[23] and Johansson and Fernström.[24]

TABLE 2 Analytical Temperature Conditions

Solvent	Temperature (°C)		
	Injector	Column	Detector
Toluene	140	130	190
Trichloroethane	110	90	120
Butanol	150	120	150

$$a^a_{uptake} = C_b \times V_{ss} + CL_b \times AUC_{(0-t)} \qquad (1)$$

where

a^a_{uptake} = the amount of solvent absorbed,
C_b = the concentration of solvent in blood at a given point in time,
V_{ss} = the apparent steady state volume of distribution,
CL_b = the clearance of the solvent, and
$AUC_{(0-t)}$ = the area under the blood concentration vs. time curve.

Assuming that (V_{ss}) and (CL_b) are constant and unaffected by the concentration in the blood, they can be assigned arbitrary values and then an arbitrary amount of absorbed solvent, a^a_{uptake}, may be calculated. The inclination of the linear portion of the invasion curve when a^a_{uptake} is plotted against time gives the arbitrary absorption rate of solvent through the skin. This is then presented as the RAR value for each solvent. The RAR values through the glove membranes, plus the skin, were then related to the value for natural rubber, which was set to 1.00.

Statistical comparison of means of the blood concentration of the exposed animals protected with glove membrane and of the unprotected control animals was carried out using Student's t-test; p values <0.05 were considered significant.

III. RESULTS

A. *In Vitro* Permeation Testing

The permeation profiles varied considerably between the different combinations of solvents and glove membranes tested (Figure 1, a to d). For butanol, no breakthrough was detected when the two thick gloves, PVC and butyl rubber, were tested. The BT (Table 3) values for toluene and 1,1,1-trichloroethane were in the same rank order for the glove materials tested, but did not correlate to their thicknesses. The permeation rate for toluene and 1,1,1-trichloroethane through butyl rubber was approximately 10,000 mg/m^2 × min and through thick PVC 6000 mg/m^2 × min at peak rate. On exposure to the natural rubber and butyl rubber gloves, the PR was reached a few minutes after BT and was maintained during the whole exposure time (Figures 1a and d). The thin PVC and the latex gloves showed similar results with low permeation rates for butanol and considerably higher rates for toluene and 1,1,1-trichloroethane (Figures 1a and b).

The permeation of butanol through the natural rubber glove and the thin PVC glove varied considerably (Figures 1a and b). In two of four test runs with the natural rubber glove, the breakthrough times were greater than 120 min.

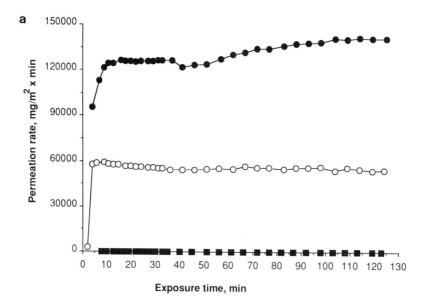

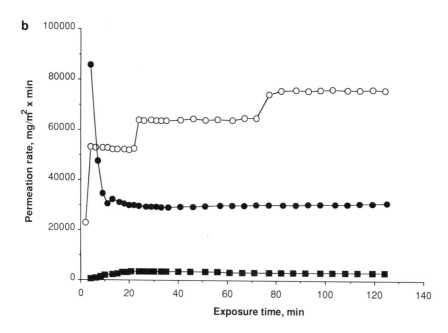

FIGURE 1 Permeation profile of solvents through (a) natural rubber gloves *(in vitro)*, (b) thin PVC gloves, (c) thick PVC gloves, and (d) butyl rubber gloves; (○—○) toluene, (●—●) 1,1,1-trichloroethane, and (■—■) butanol. (From Mellström, G. and Boman, A., *Contact Derm.*, 26, 120, 1992. With permission.)

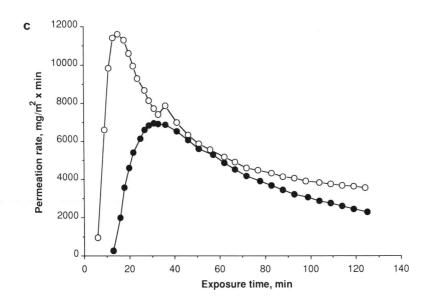

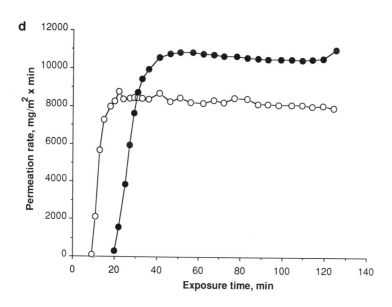

FIGURE 1 (continued).

TABLE 3   Breakthrough Time (BT, in Min) From *In Vitro* and *In Vivo* Assessment of Protective Gloves

	Solvents					
	TOL		TCE		BUT	
Glove materials	*In vitro*	*In vivo*	*In vitro*	*In vivo*	*In vitro*	*In vivo*
PVC, thick	6–11	<10	13–22	<20–<30	>120	>360
Butyl rubber	9–10	<10	20–25	<10–<60	>120	>360
Natural rubber	2–4	<10	4–5	<10	8–>120	<10–<210
PVC, thin	2	<10	4–5	<10	4	<10–<40

From Mellström, G. and Boman, A., *Contact Derm.*, 26, 120, 1992. With permission.

The results from calculations of the RPR are exemplified for 1,1,1-trichloroethane in Figure 2 and are summarized for all solvents in Tables 4 and 5.

B. *In Vivo* Testing

Exposure of normal skin to toluene and 1,1,1-trichloroethane resulted in an immediate increase in the blood concentration of the solvents, followed by a decline and setting a steady-state concentration lower than during the first 60 min (Figure 3). The introduction of a glove membrane between the solvent and skin reduced the blood concentration of the solvents. The most pronounced difference in concentration between protected and unprotected skin in the control animals was seen during the initial 60 min of the exposure. This difference was seen for all types of membranes investigated. For butyl rubber the reduction was statistically significant throughout the whole exposure period (Figure 4). Natural rubber gave a significant reduction during 90 min.

The exposure to butanol was carried out with normal, untreated skin and with stripped skin (Figures 5 and 6). Stripping increased the absorption dramatically. If, however, stripped skin was protected with butyl rubber or thick PVC membrane during exposure, the concentration of solvent was significantly reduced and was below those of normal skin. No breakthrough was detected when intact skin was exposed covered with the two thickest gloves of PVC and butyl rubber, but on stripped skin measurable amounts of breakthrough occurred and these figures were used for the assessment (Table 5).

Recalculation of the blood concentration data and plotting vs. time as relative absorption rate (RAR) is exemplified for trichloroethane in Figure 7. Results of BT and the relative permeation and absorption rates from the *in vitro* and *in vivo* testing are summarized in Tables 3, 4, and 5.

Percutaneous Absorption Studies in Animals 99

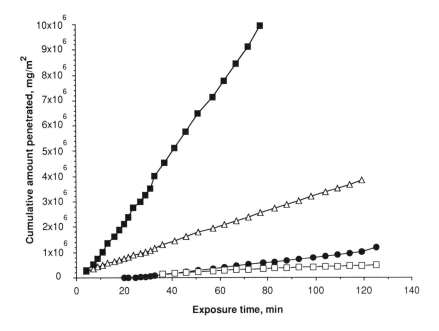

FIGURE 2 Relative permeation rate *(in vitro)* for 1,1,1-trichloroethane through protective gloves; (□—□) thick PVC, (●—●) butyl rubber, (■—■) natural rubber, (Δ—Δ) thin PVC. (From Mellström, G. and Boman, A., *Contact Derm.*, 26, 120, 1992. With permission.)

TABLE 4 Calculated Figures For Relative Permeation (RPR) and Relative Absorption (RAR) Rates for Toluene and 1,1,1-Trichloroethane from *In Vitro* and *In Vivo* Assessment of Gloves

	Solvents			
	TOL		TCE	
Glove materials	RPR	RAR	RPR	RAR
PVC, thick	0.08	1.42	0.03	1.04
Butyl rubber	0.15	0.52	0.08	0.44
Natural rubber[a]	1.00	1.00	1.00	1.00
PVC, thin	1.26	1.17	0.23	1.23
Unprotected skin	—	1.78	—	1.45

[a] The figure for natural rubber was set to 1.00 (see text).
From Mellström, G. and Boman, A., *Contact Derm.*, 26, 120, 1992. With permission.

TABLE 5 Calculated Figures for RPR and RAR for Butanol from *In Vitro* and *In Vivo* Assessment of Gloves

Glove materials	RPR	RAR	RAR(S)
PVC, thick	ND	ND	1.05
Butyl rubber	ND	ND	0.66
Natural rubber[a]	1.00	1.00	—
PVC, thin	47.0	2.2	—
Unprotected skin	—	5.57	—

Note: ND = not detected, S = stripped skin.

[a] The figure for natural rubber was set to 1.00 (see text).
From Mellström, G. and Boman, A., *Contact Derm.*, 26, 120, 1992. With permission.

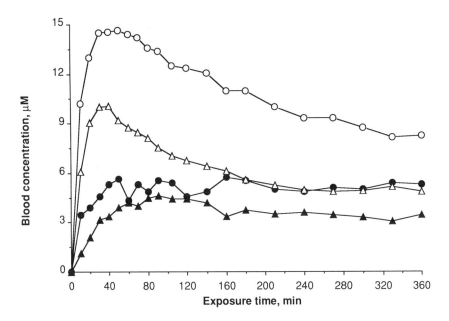

FIGURE 3 Absorption profile *in vivo* of solvents through protected and unprotected skin; (○—○) toluene, unprotected skin; (●—●) toluene, natural rubber protected skin; (△—△) 1,1,1-trichloroethane, unprotected skin; (▲—▲) 1,1,1-trichloroethane, natural rubber protected skin. (From Boman, A. and Mellström, G., *Contact Derm.*, 21, 260, 1989. With permission.)

IV. DISCUSSION

When assessing the protective effects of glove materials against skin absorption of chemicals animal models are a useful complement to a purely technical standard method such as the ASTM F 739-85[8] or other methods.[9] In a technical permeation study *in vitro* the solvent goes through a single glove membrane in

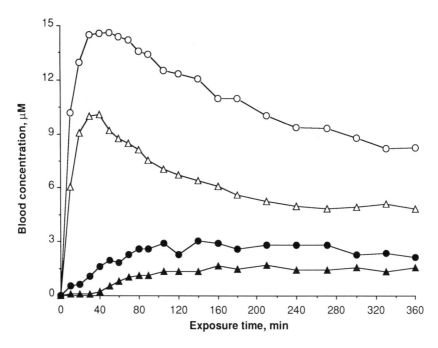

FIGURE 4 Absorption profile *in vivo* in guinea pigs of solvents through protected and unprotected skin. (○—○) toluene, unprotected skin; (●—●) toluene, butyl protected skin; (△—△) 1,1,1-trichloroethane, unprotected skin; (▲—▲) 1,1,1-trichloroethane, butyl rubber protected skin. (From Boman, A. and Mellström, G., *Contact Derm.*, 21, 260, 1989. With permission.)

a two-chamber test cell in an open system at room temperature. On the other hand, in an *in vivo* study, to which these *in vitro* results are compared, the solvent permeates through two barriers — glove and skin — and is based on measurements in guinea pig blood, a closed circulation system, and at a higher temperature. The biological method also takes into account factors influencing absorption that are not considered in the technical testing, such as local effects on the skin of the permeating chemical influencing the absorption, or of the protective material separately or in combination. Occlusion, as provided by the glove membrane, is considered to be one of the most crucial factors in increasing percutaneous absorption.[10] However, as the two test methods differ in performance and test temperatures, the results can only be compared on a relative basis and rank order.

In the *in vitro* testing, representatives for four different types (A to D) of permeation behavior were described by Nelson et al.[25] as being caused by interaction between the test chemical and the protective glove materials during the permeation testing procedure. The most typical pattern observed was the A type, where the permeation rate stabilizes at a "steady state" level (Figures 1 and 4). Type B (Figure 2), where the permeation behavior is due to the material

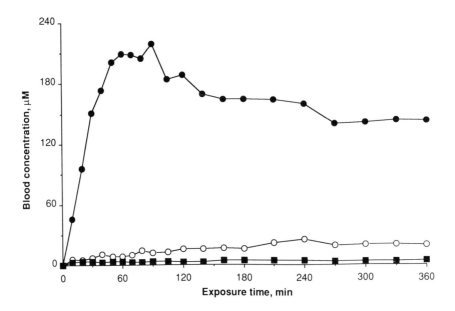

FIGURE 5 Absorption profile *in vivo* in guinea pigs of butanol through protected, unprotected stripped, and unprotected normal skin; (○—○) unprotected normal skin, (■—■) butyl rubber or thick PVC protected stripped skin, (●—●) unprotected stripped skin. (From Mellström, G. and Boman, A., *Contact Derm.*, 26, 120, 1992. With permission.)

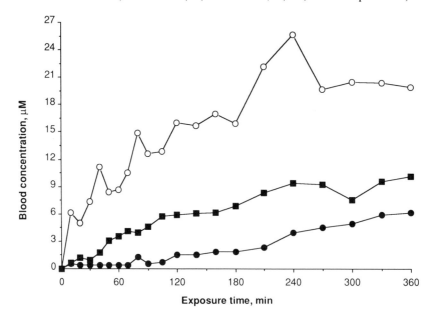

FIGURE 6 Absorption profile *in vivo* in guinea pigs of butanol through protected and unprotected skin; (○—○) unprotected skin, (●—●) thin PVC protected skin, (■—■) natural rubber protected skin.

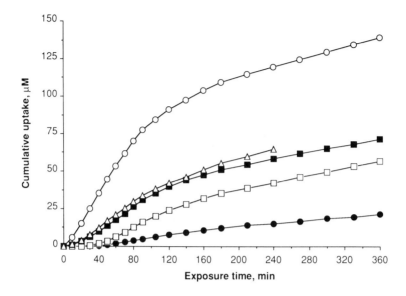

FIGURE 7 Relative absorption rate *in vivo* for 1,1,1-trichloroethane through protective gloves; (□—□) thick PVC, (●—●) butyl rubber, (■—■) natural rubber, (△—△) thin PVC, (○—○) unprotected skin.

specimen being structurally modified by the test chemical, resulting in a change in permeation rate; and type D behavior, with an initial high rate decreasing slowly during the exposure time, occurs when there is moderate to heavy swelling of the material (Figure 2).

If the glove membranes are structurally modified by the solvent, changes in permeation behavior will be more evident during a long exposure time and will influence the test results as noted with PVC, toluene, and trichloroethane.

It would be reasonable to expect the BT values to be in the same rank order in both kinds of investigation for the different solvents, and that the RPR and RAR values would vary. The BT values in Table 3 are presented as ranges, or as less (<) or greater than (>), since a mean value would not give proper information in cases of great variation. Despite this, the BT values from the *in vitro* and *in vivo* testing corresponded well and were in the same rank order for all three solvents.

The RPR values for toluene fluctuated considerably through the different materials but followed the same rank order as the thickness of the gloves (Table 4). Permeation of toluene and trichloroethane through the butyl rubber and natural rubber membranes showed the most typical pattern (type A). The RAR value for toluene through thick PVC glove was the highest, i.e., this glove had the least protective effect. Still, these gloves reduced skin absorption by about 50% during the first 60 min. Trichloroethane showed the highest RPR for the natural rubber glove, with considerably lower rates for the two thick gloves tested. For trichloroethane, however, the protective effect of the thick PVC glove and the natural rubber glove was almost the same as in the *in vivo* test (Table 4). The RPR values for butanol disclose the greatest differences. For the natural rubber and thin PVC gloves, a 47-fold difference in the *in vitro* testing is reduced to a 2-fold difference in the *in vivo* test system (Table 5).

For the two thick gloves, PVC and butyl rubber, no BT was detected for butanol in either of the test methods, i.e., the permeation and absorption of the solvent were so low that the concentration in the gas flow *(in vitro)* and in the blood *(in vivo)*, respectively, were below the MDL. The *in vivo* results show a reduction in blood concentrations in animals exposed to solvents in conjunction with glove membranes. The reduction was most pronounced during the first 1 to 2 h of exposure. For one glove membrane (natural rubber), protection was not satisfactory, the blood analyses showing detectable amounts for all three solvents tested even in the first samples (after 10 min) (Figures 6 and 9).

The permeation of a solvent through a glove membrane and/or the skin is assumed to be a passive diffusion process following Ficks' first law of diffusion:[22,27]

$$J = -D\partial C/\partial x \qquad (2)$$

where

- J is the rate of transfer per unit area,
- C is the concentration of the diffusing chemical,
- x is the thickness of the barrier, and
- D is the diffusion coefficient, which is assumed to be neither a function of distance nor of concentration.

The glove membranes are exposed to the liquid phase of the solvent, but if the glove membrane is without imperfections, e.g., pinholes, the skin will be more exposed to the saturated vapor of the solvent, and under occlusive conditions.

Although the PVC and natural rubber gloves were permeated after a very short time they did show a reduction in blood concentration of solvent and may thus be used as gloves for short exposure times, such as in splash situations or where the appliance of finger tactility in the work situation does not allow the use of thick gloves. However, the gloves should be replaced immediately after contact.

Skin forms a good natural barrier against chemical and other agents as long as it is intact. If the barrier is damaged in some way due to skin diseases or small wounds, the protective effect can be reduced and hydrophilic solvents in very low concentration can permeate.[23] This was demonstrated for butanol, where the stripped skin protected with PVC or butyl rubber gloves was obviously permeated by butanol in a concentration below the MDL. Butanol at the concentration used was absorbed through stripped skin, but not through intact

TABLE 6 Ranking of Protective Effects Comparing *In Vitro* and *In Vivo* Testing

Glove Materials	Solvents								
	TOL			TCE			BUT		
	BT	RPR	RAR	BT	RPR	RAR	BT	RPR	RAR
PVC, thick	2	1	3	2	1	3	1	1	1
Butyl rubber	1	2	1	1	2	1	1	1	1
Natural rubber	3	3	2	3	4	2	2	2	2
PVC, thin	4	4	4	4	3	4	3	3	3

Note: 1 is most protective, 4 is least protective.
From Mellström, G. and Boman, A., *Contact Derm.*, 26, 120, 1992. With permission.

skin (Table 5). This should be kept in mind, as protective gloves are usually recommended for protecting hands with current contact dermatitis.

A comparison of the *in vitro* permeation behavior type and the *in vivo* test results suggests that a type A permeation curve gives lower RAR values. This may be because the membranes are not, or are only slightly, structurally affected by the solvent. However, if the materials have been structurally affected, which is the case for the B- and D-type permeation curves, the membranes have become swollen and wrinkled (increased area). A membrane, once permeated or penetrated by a chemical, may therefore act as an absorption-promoting factor by occluding the skin and thus potentiating the absorption of the chemical.

The rank order of the BT figures for toluene and trichloroethane does not correspond to the thicknesses of the gloves (Table 6). This indicates that the material itself, its structure and quality, have a greater influence on the protective effect than the thickness. After breakthrough, the thickness is of minor importance and the value of the diffusion coefficient will influence subsequent permeation/absorption of the solvents. It is only for butanol that the RAR values and RPR values of the solvents are in the same rank order and related to the thicknesses of the glove materials. For toluene and trichloroethane, the protective effect had the same rank order only for the BT figures, not for RPR or RAR figures. Thus, thickness cannot generally be used as a criterion for selection.

Breakthrough time values from *in vitro* testing — in most cases the only available permeation data — provide the best available guide in the selection of protective gloves and how long they can be used during continuous exposure to chemicals. This investigation also shows that a protective glove with relatively high permeation rate values in an *in vitro* testing system can perform as a barrier, reducing the absorption rate in an *in vivo* situation under certain circumstances.

Obviously, not every choice of gloves in the industry can be based on an *in vivo* assessment of the glove materials available on the market. But manufacturers' development of new materials and reviews of older materials should, in addition to purely technical testing, include an *in vivo* assessment. Although the animal model described does not give absolute absorption rates of solvents, it is suitable for an *in vivo* assessment of glove materials already available on the market, and can also be used for predictive testing of new products as it discerns glove materials well. However, it must be kept in mind that the most reliable data will be obtained from human experiments, and preferably in real work situations.[13,14]

V. CONCLUSION

A general conclusion is that when protection against skin exposure to organic solvents is needed, and no other means such as altered handling or altered working routines are possible, gloves are the best choice. However, care must be taken to make a well considered selection among the various materials available on the market and to remember that protection is relative and limited in time.

REFERENCES

1. Wahlberg, J. E., Edema-inducing effects of solvents following topical administration, *Dermatosen,* 32, 91, 1984.
2. Wahlberg, J. E., Erythema-inducing effects of solvents following epicutaneous administration to man — studied by laser Doppler flowmetry, *Scand. J. Work Environ. Health,* 10, 159, 1984.
3. Kronevi, T., Wahlberg, J. E., and Holmberg, B., Skin pathology following epicutaneous exposure to seven organic solvents, *Int. J. Tiss. Reactions,* 3, 21, 1981.
4. Kronevi, T., Wahlberg, J. E., and Holmberg, B., Histopathology of skin, liver, and kidney after epicutaneous administration of five industrial solvents to guinea pigs, *Environ. Res.,* 19, 56, 1979.
5. Wahlberg, J. E. and Boman, A., Comparative percutaneous toxicity of ten industrial solvents in the guinea pig, *Scand. J. Work Environ. Health,* 5, 345, 1979.
6. Mellström, G., Protective effect of gloves — compiled in a database, *Contact Derm.,* 13, 162, 1985.
7. Mellström, G. A., Lindahl, G., and Wahlberg, J. E., DAISY: reference database on protective gloves, *Sem. Dermatol.,* 8, 75, 1989.
8. ASTM Standard Method, *Test for Resistance of Protective Clothing Materials to Permeation By Hazardous Liquid Chemicals,* F 739-85, American Society for Testing and Materials, Philadelphia, PA, 1986, 426.
9. International Organization for Standardization (ISO), Protective clothing. Protection against liquid chemicals. Determination of resistance of air-impermeable materials to permeation by liquids, Draft International Standard ISO/DIS 6529.4, (CEN TC162 WG3/N20, Chemical Protection), 1988.
10. Bucks, D. A. W., Maibach, H. I., and Guy, R. H., Occlusion does not uniformly enhance penetration *in vivo,* in *Percutaneous Absorption,* Bronaugh, R. L. and Maibach, H. I., Eds., Marcel Dekker, New York, 1985, 77.

11. Boman, A. and Mellström, G., Percutaneous absorption of three organic solvents in the guinea pig. IV. Effect of protective gloves, *Contact Derm.,* 21, 260, 1989.
12. Hogstedt, C. and Ståhl, R., Skin absorption and protective gloves in dynamite work, *Am. Ind. Hyg. Assoc. J.,* 41, 367, 1980.
13. Lauwerys, R. R., Kivits, A., Lhoir, M., Rigolet, P., Houbeau, D., Buchet, J.-P., and Roels, H., A biological surveillance of workers exposed to dimethylformamide and influence of skin protection on its percutaneous absorption, *Int. Arch. Occup. Environ. Health,* 45, 189, 1980.
14. Brooks, S. M., Anderson, L., Emmett, E., Carson, A., Tsay, J. Y., Elia, V., Buncher, R., and Karbowsky, R., The effects of protective equipment on styrene exposure in workers in the reinforced plastics industry, *Arch. Environ. Health,* 35, 287, 1980.
15. Lidén, C., Occupational dermatoses in a film laboratory, *Contact Derm.,* 10, 77, 1984.
16. Moursiden, H. T. and Faber, O., Penetration of protective gloves by allergens and irritants, *Trans. St. Johns Hosp. Dermatol. Soc.,* 59, 230, 1973.
17. Pegum, J. S., and Medhurst F. A., Contact dermatitis from penetration of rubber gloves by acrylic monomer, *Br. Med. J.,* 2, 141, 1971.
18. Pegum, J. S., Penetration of protective gloves by epoxy resin, *Contact Derm.,* 5, 281, 1979.
19. Wall, L. M., Nickel penetration through rubber gloves, *Contact Derm.,* 6, 461, 1980.
20. Nigg, H. N. and Stamper, J. H., Field evaluation of protective clothing. Experimental designs, *Performance of Protective Clothing: Second Symposium, ASTM STP 989,* Mansdorf, S. Z., Sager, R., and Nielsen, A. P., Eds., American Society for Testing and Materials, Philadelphia, PA, 1988, 776.
21. Schwope, A. D., Costas, P. P., Stull, J. O., and Weitzman, D. J., *Guidelines for Selection of Protective Clothing,* 3rd ed., Vol. II and III, Am. Conf. Gov. Ind. Hyg., Cincinnati, OH, prepared by Arthur D. Little Inc., Cambridge, MA, 1987.
22. Jamke, R. A., Understanding and using chemical protective clothing, *Chemical Protective Clothing Performance in Chemical Emergency Response, ASTM STP 1037,* Perkins, J. L., Stull, J. O., Eds., American Society for Testing and Materials, Philadelphia, 1989, 11.
23. Boman, A., Factors influencing the percutaneous absorption of organic solvents. An experimental study in the guinea pig, *Arbete och Hälsa 1989:11,* Arbetsmiljöinstitutet, Solna, Sweden, 1989.
24. Johansson, G. and Fernström, P., Percutaneous uptake rate of 2-butoxyethanol in the guinea pig, *Scand. J. Work Environ. Health,* 12, 499, 1986.
25. Nelson, G. O., Lum, B. Y., Carlson, G. J., Wong, C. M., and Johnson, J. S., Glove permeation by organic solvents, *Am. Ind. Hyg. Assoc. J.,* 42, 217, 1981.
26. Dugard, P. H., Skin permeability theory in relation to measurements of percutaneous absorption in toxicology, in *Dermatoxicology,* 3rd ed., Marzulli, F. N. and Maibach, H. I., Eds., Hemisphere, Washington, D.C., 1987, 95.
27. Schwope, A. D., Goydan, R., Reid, R. C., and Krishnamurthy, S., State-of-the-art review of permeation testing and the interpretation of its results, *Am. Ind. Hyg. Assoc. J.,* 49, 557, 1988.

9

Standard Quality Control Testing and Virus Penetration

C. David Lytle, W. Howard Cyr, Ronald F. Carey,
David G. Shombert, Bruce A. Herman,
James G. Dillon, Leroy W. Schroeder,
Harry F. Bushar, and Helen J. Rosen Kotilainen

TABLE OF CONTENTS

  I. Introduction 110
 II. Standard Tests for Glove Integrity 110
     A. The FDA 1000 ml Water Leak Test 110
     B. Comparisons of Standard Tests 112
        1. An Interlaboratory Comparison 112
        2. Sensitivities of Some Tests 113
III. Penetration by Small Particles Through Undetected Pinholes 116
     A. Testing With Viral-Sized Particles 117
     B. Testing With Viruses 118
        1. Important Properties of Virus Particles 118
        2. Virus Penetration Through Used and
           Intact Gloves — A Review 118
        3. Viral Penetration Through Punctures in Gloves 121
 IV. Material Integrity 123
     A. Material Fatigue 123
     B. Other Factors That Influence Integrity 124
 IV. Summary 125
References 125

I. INTRODUCTION

The recent concern about the quality of medical gloves has arisen from the possibility of transmission of a deadly disease: Acquired Immune Deficiency Syndrome (AIDS). Because of the AIDS epidemic, there have been a number of research studies on barrier protection for the health care worker, with primary concern about barriers to the human immunodeficiency virus (HIV), the suspected etiological agent of AIDS. These studies have also provided data useful to the prevention of other viral diseases, such as hepatitis B (HBV) and herpes simplex (HSV).

The original practice of wearing gloves for surgery started in an effort to reduce postoperative infections among the surgical patients. The use of gloves has expanded outside the surgical suites, and is now considered a primary precaution for protection by nearly all health care workers. The primary concern is the integrity of the glove as a barrier to transmission of infectious agents, especially viruses.

This chapter will focus on the tests for medical glove integrity and the relevance of these tests in assuring barrier effectiveness to virus transmission. In addition, information on perforation of gloves during use is presented.

The passage of viruses through gloves will generally occur when defects, e.g., pinholes or tears, are present. This passage is called penetration. On the other hand, permeation — where a substance might migrate or diffuse through intact fabric — is unlikely for viruses unless the glove is greatly stretched or is degraded by certain chemicals.[1] Hence, the term penetration will be used throughout this chapter in the above-described sense. The word "leakage" is defined as the passage of water (or any challenge fluid) through the glove, and could result from penetration or permeation.

II. STANDARD TESTS FOR GLOVE INTEGRITY

A. The FDA 1000 ml Water Leak Test

The U.S. Food and Drug Administration (FDA) utilizes a quality assurance test known as the "1000 ml water leak test". The protocol for this test, as described in Chapter 21 of the Code of Federal Regulations (21 CFR 800.20), is to fasten (using elastic strapping with Velcro® or other fastening material) a glove to a (clear) glass or plastic cylinder (2.75 in. in diameter by 15 in. long), using 1.5 in. of the cuff of the glove as the attachment area. The plastic cylinder serves as a fill tube; with some gloves the water level may extend several inches up the fill tube. The test setup is shown in Figure 1. The American Society for Testing and Materials (ASTM) now also has a 1000 ml water leak test for medical gloves.

Any visually detectable water on the outside of the glove, i.e., a leak within a 2-min observation time, is considered a failure. Leaks within 1.5 in. of the cuff are disregarded. While this test cannot detect microscopic holes, it is

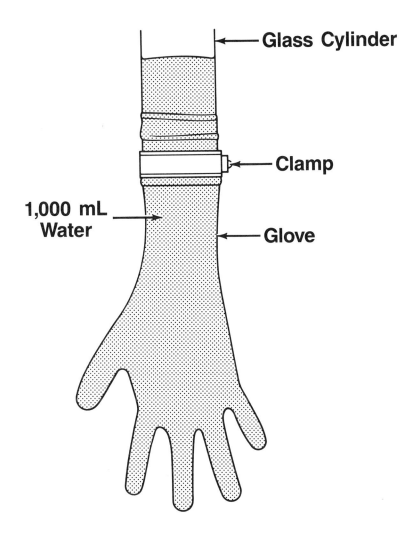

FIGURE 1 Diagram illustrating the setup for the FDA 1000 ml water leak test.

intended to indicate overall manufacturing quality. The regulation requires that 97.5% of surgical gloves and 96% of exam gloves pass this test (see below). Manufacturers routinely test to these specifications, and FDA inspects manufacturers' records and samples their production lots to assure compliance.

FDA collects samples from the manufactured lots of gloves and tests them according to the size of the lot, using a statistical sampling scheme which is described in the Code of Federal Regulations (21 CFR 800.20). The final rule, published in the December 12, 1990 *Federal Register* and effective March 12,

1991, defined the acceptable quality level (AQL) at 2.5% for surgeons' gloves and at 4.0% for patient examination gloves. For regulatory purposes the FDA refers to this as the adulteration level. FDA collects random samples from lots of medical gloves and performs the 1000 ml water leak test in accordance with FDA's sampling inspection plans, which have been derived from the tables in MIL-STD-105E, the military standard for "Sampling Procedures and Tables for Inspection by Attributes", May 10, 1989. FDA uses single sampling for lots of ≤1200 gloves and multiple sampling for lots of >1200 gloves, where the term "lot" means a collection of gloves from which a sample is to be drawn and inspected and may differ from a collection of gloves designated as a lot for other purposes, e.g., production or shipment.

Glove lots that are tested and rejected when using the 1000 ml water leak test and the AQLs specified in the final rule are considered adulterated and subject to regulatory action, such as detention of imported gloves or seizure of domestic gloves. The actual sampling plans (sample size and accept/reject numbers) are shown in Tables 1 and 2. Note that sampling and testing may cease when a lot is determined to be violative, i.e., rejected.

B. Comparisons of Standard Tests

1. An Interlaboratory Comparison

Douglas et al.[1] reported on a "round-robin" interlaboratory comparison of three standard test methods using four types of medical gloves: sterile latex surgical, latex nonsterile, vinyl nonsterile, and film (copolymer) nonsterile gloves. The tests for penetration included a chemical fill test, a water fill test, and an air inflation test.

In the chemical fill test, the outside of the glove was dipped in 2% phenolphthalein and allowed to dry. The inside of the glove was then filled with a 2% carbonate solution. If any carbonate leaked, a pink color change could be detected on the outside of the glove. This test method was the most sensitive, overall, with latex gloves, but gave results suggestive of permeation of the carbonate (rather than penetration through pinholes) for the vinyl and plastic gloves. Furthermore, the authors reported that this test was not cost effective. The water fill test consisted of filling the glove with water to a level of 30 cm above the top of the middle finger (thus employing a constant pressure head from glove to glove, rather than the constant challenge volume of the FDA water fill test), and checking for leakage within 5 min. This test was slightly less sensitive than the chemical test, was cost effective, and was reproducible from lab to lab. The air inflation test (inflation with air to a volume 350% greater than the original volume of the glove) consisted of a visual examination of the inflated glove for holes, tears, thin spots, and imbedded foreign material. This inflation test was less sensitive than either the chemical test or water fill test.

TABLE 1 Sampling Scheme for Lots of Surgeons' Gloves Where the Accepted Quality Level is 2.5%

Lot size	Sample	Sample size	Number examined	Number defective Accept	Number defective Reject
>35,000	1st	125	125	2	9
	2nd	125	250	7	14
	3rd	125	375	13	19
	4th	125	500	19	25
	5th	125	625	25	29
	6th	125	750	31	33
	7th	125	875	37	38
10,001–35,000	1st	80	80	1	7
	2nd	80	160	4	10
	3rd	80	240	8	13
	4th	80	320	12	17
	5th	80	400	17	20
	6th	80	480	21	23
	7th	80	560	25	26
3,201–10,000	1st	50	50	0	5
	2nd	50	100	3	8
	3rd	50	150	6	10
	4th	50	200	8	13
	5th	50	250	11	15
	6th	50	300	14	17
	7th	50	350	18	19
1,201–3,200	1st	32	32	0	4
	2nd	32	64	1	6
	3rd	32	96	3	8
	4th	32	128	5	10
	5th	32	160	7	11
	6th	32	192	10	12
	7th	32	224	13	14
501–1,200	Single	80	80	5	6
281–500	Single	50	50	3	4
151–280	Single	32	32	2	3
51–150	Single	20	20	1	2
5–50	Single	5	5	0	1

The authors concluded that the water fill test was most useful. However, they found that the constant pressure head was not convenient because of glove to glove variation; they suggested that the constant fill volume method, as used by FDA and others, should be more convenient.

2. Sensitivities of Some Tests

Analysis of several leakage tests for surgical and examination gloves has been reported by Carey et al.[2] A total of 6 documented test protocols were evaluated, using samples from 16 glove manufacturers. All of these tests involve detection of the passage of air or water through the glove barrier (see Table 3).

TABLE 2 Sampling Scheme for Lots of Patient Examination Gloves Where the Accepted Quality Level is 4.0%

Lot size	Sample	Sample size	Number examined	Number defective Accept	Number defective Reject
>10,000	1st	80	80	2	9
	2nd	80	160	7	14
	3rd	80	240	13	19
	4th	80	320	19	25
	5th	80	400	25	29
	6th	80	480	31	33
	7th	80	560	37	38
3,201–10,000	1st	50	50	1	7
	2nd	50	100	4	10
	3rd	50	150	8	13
	4th	50	200	12	17
	5th	50	250	17	20
	6th	50	300	21	23
	7th	50	350	25	26
1,201–3,200	1st	32	32	0	5
	2nd	32	64	3	8
	3rd	32	96	6	10
	4th	32	128	8	13
	5th	32	160	11	15
	6th	32	192	14	17
	7th	32	224	18	19
501–1,200	Single	80	80	7	8
281–500	Single	50	50	5	6
151–280	Single	32	32	3	4
91–150	Single	20	20	1	2
26–90	Single	13	13	1	2
3–25	Single	3	3	0	1

Mechanical, electrical, and optical techniques for inducing submillimeter holes were developed and employed for creating test holes. The prescribed tests were performed and the results compared to calculations of expected sensitivities.

For a given glove hole (characterized by its diameter, the glove thickness at the point of the hole, and the glove material) and a given test method (characterized by the test pressure across the glove surface and by the test fluids), detectability of a hole may be limited by either surface tension effects or by the flow rate of the leaking fluid (water or air) through the hole.

For five of the six protocols, surface tension effects at the water/air interface control the sensitivity of the test. For water/air interfaces, the water pressure, P (mmHg), necessary to balance the surface tension is given by:

$$P \approx 1.1/R \tag{1}$$

TABLE 3 Sensitivities of Standardized Tests to Determine Quality of Gloves

Glove type	Test	Description of test	Best[a] sensitivity (µm)	Worst sensitivity (µm)
Vinyl exam	MIL-G-36393 & MIL-G-36593	300 ml water fill, 5 min	50	No limit
Latex exam	MIL-G-36724A	20 mmHg air pressure under water	Not usable	
Latex exam	ASTM D3578-77	300 ml water fill, 2 min	100	No limit
Latex surgery	MIL-G-36596A	20 mmHg air pressure under water	25	70
Latex surgery	MIL-G-36596A	12–14 mmHg air deflation (in air)	25	100
Latex	ASTM D3577-78a	11 mmHg air pressure under water	75	450
All	FDA	1000 ml water fill, 2 min	25[b]	50[b]

[a] Size of hole detected by test.
[b] An estimate based upon roughly doubling the height of the water in the glove when increasing the volume of fill from 300 to 1000 ml.

where R (mm) is the radius of curvature of the water surface as the water advances through the hole or as it accumulates on the surface of the glove. The water pressure necessary to produce a detectable leak must exceed the value of P for the smallest radius of curvature experienced as the water moves into and out of the hole. Theoretically, the critical pressure depends upon the hole size and on the wettability of the glove material by water. For the gloves examined in this study, the smallest radius of curvature was approximately given by the radius of the hole. This criterion also applies to an air bubble attempting to pass into a water bath. For the maximum water or air pressure produced by these standard tests, 20 mmHg, the smallest hole for which surface tension can be overcome is one with diameter $d \simeq 0.11$ mm. In practice, other factors such as handling of the glove, or variations in the powders used on the glove, or stretching of the glove by the test conditions, will allow somewhat smaller holes to be observed.

The water flow rate, Q (in microliters per second), through a glove hole, if the surface tension effects have been overcome, is limited by Poiseuille's equation, which, for the conditions of these tests (room temperature, where the viscosity is 1.0 cP) can be written as:

$$Q \approx 54 \, P \, d^4 \qquad (2)$$

For example, if $d = 0.05$ mm and $P = 20$ mmHg, then $Q = 6.8$ μl/s which implies that in just 15 s, 100 μl comes through the hole. (Approximately 0.3 μl can be visually detected.) The implication is that the flow rate will not limit detection of a leak when leakage requires passage of water into air or air into water; i.e., at the pressures produced by these tests, holes large enough to allow surface tension effects to be overcome will allow passage of easily detected volumes of fluid.

For one of these tests, the air deflation protocol under MIL-G-36596A, surface tension effects are not relevant. Here, Poiseuille flow of air can be written as: $Q \approx 3500\, d^4$. However, detection depends upon a change in pressure, and volume/pressure relationships for gloves are highly variable from glove to glove for a given pressure and highly pressure dependent for a given glove. Therefore, theoretical sensitivities for this test can not be easily estimated.

Experimental results for the six tests are summarized in Table 3. In general, there was reasonable agreement between calculation and measurement (within about a factor of 2), although there were some exceptions. Minimum detectable hole sizes, before the holes were stretched by the tests, were generally in the range of 0.025 to 0.10 mm in diameter. Because of inherent limitations of the test protocols, it is clear that much larger holes can go undetected under certain circumstances, especially in examination gloves. For example, the 300-ml fill fails to completely fill many of these gloves, leaving significant portions of the barrier untested. One test, MIL-G-36724A, was impossible to perform, since it specified a 20 mmHg air pressure for gloves which cannot be inflated beyond about 12 mmHg. All tests had inherent limitations resulting in sensitivities that varied from glove to glove, or that were quite nonuniform over the surface of the glove being tested, or both.

FDA has chosen a 1000 ml water fill test as the standard test for quality assurance for both examination and surgical gloves. A comparable experimental study for this test method has not been done, although one could estimate that its best sensitivity (minimum detectable hole size anywhere on a glove) would be roughly in the range of 0.025 mm, and its worst sensitivity (minimum detectable hole size at the point of lowest sensitivity on the glove) is perhaps double that.

III. PENETRATION BY SMALL PARTICLES THROUGH UNDETECTED PINHOLES

The biological entities of greatest concern are viruses, primarily because they are the smallest human pathogens readily transmitted by body fluids, especially blood. For laboratory tests, it is clearly desirable for safety reasons to use substitutes for the more dangerous human viruses. Three choices have been used in testing the barrier properties of some barrier materials: viral-sized fluorescent beads, bacteriophage, and human viruses which would cause less safety concerns than HIV.

A. Testing With Viral-Sized Particles

There are no published reports using viral-sized particles to evaluate the effectiveness of gloves as barriers to HIV or other viruses. However, Retta et al.[3] developed a test for condoms which could be adapted to gloves. The principle of the test is to challenge the barrier under conditions which simulate physiologic conditions, particularly with low surface tension liquids on both sides of the barrier, so that even holes of viral size will wet and potentially allow passage of viral-sized particles.

Particle size, pH, surface tension, viscosity, temperature, pressure, geometry, and time were considered in the condom test system. A suspension of polystyrene microspheres (110 nm in diameter) labeled with fluorescent dye (Fluoresbrite®, Polysciences, Inc., Warrington, PA) simulated viruses. The aqueous suspension was diluted with phosphate buffered saline to a concentration of 10^{11} microspheres per milliliter or more, to yield sufficient detection sensitivity. Because the initiation of fluid flow through small pores is dependent on surface tension effects, a surfactant was added to the suspension to reduce the surface tension to 0.03 N/m^2 (vs. 0.072 N/m^2 for water). The contact angle of the fluid with the pores in the latex was then consistently less than 90°, which assured that surface tension did not pose a barrier to flow initiation. The viscosity of the suspension at 37°C was measured to be 0.7 cP. The flow, Q (in microliters per second), through any glove pore would then be given by the Poiseuille equation as:

$$Q \approx 31 P d^4 \qquad (3)$$

for transmembrane pressure P (mmHg) and pore diameter d (mm). Transmembrane pressures during glove use, however, are difficult to determine. An estimate of the worst case has been done[29] by measuring the pressure generated at the surface of a scalpel by a gloved hand. Pressures as high as 300 mmHg were found. This value would provide maximum test sensitivity without exceeding physiologic pressure, but any glove tested at this pressure would need to be restrained. The fluorescence measurement can detect about 1 µl of transferred fluid.

For any test which detects barrier defects in the range of viral dimensions, knowledge of effective particle sizes is an important issue. Electron microscopy was used to view the microspheres used for condom tests. Filtration experiments confirmed that the fluorescent label was not present in significant amounts in a form dissociated from the microspheres. However, the amount of this unbound label increased with exposure to surfactant and with exposure to latex.[30] Nor can the possibility of other differences between particles and viruses, such as electrostatic charge and geometric shape, be eliminated from having influence upon interpretation of results.

B. Testing With Viruses

1. Important Properties of Virus Particles

Viruses are made up of nucleic acid covered by protein coats that, in some cases, are enclosed in lipid-containing envelopes. Studies of the filtration of viruses through well-defined holes have indicated that the filtration sizes of many viruses were essentially the same as those seen in electron micrographs,[4,5] indicating that many approximately spherical virus particles were not deformed significantly when passing through small holes, did not have Stokes radii (effective hydraulic radii) significantly larger than their physical radii, and therefore could be regarded as solid particles. Furthermore, passage through a hole is a physical phenomenon that is not related to the disease potential of the virus. In addition to size and shape, another important property of a virus particle may be its electrical charge, which depends primarily on the nature of the proteins on the particle surface. In addition, the chemical properties may be important in the event of possible chemical interaction, e.g., with viricidal agents. Surrogate viruses may be used to substitute for pathogenic viruses in laboratory tests if adequate matches can be made for these physical parameters.[6]

The shapes of nearly all animal viruses are approximately spherical. This includes the human viruses that cause AIDS (retroviruses), herpes simplex (herpesviruses), papillomas (papilloma viruses), polio (picornaviruses), and hepatitis B. Several families of bacterial viruses (also called bacteriophage) also have nearly spherical shapes and sizes close to those of the animal viruses (see Table 4). Bacterial viruses are useful as surrogate viruses for the human pathogens, not only because they are inherently safer, but because their culture characteristics allow faster and less expensive experimental protocols. In order to demonstrate that a barrier material is adequate to prevent penetration by selected human viruses, one could substitute a bacterial virus, such as ϕX174, which is smaller than the human viruses of concern.

In actual laboratory tests of virus penetration through materials, the electrical or chemical properties of the virus particle may also be important. For example, virus particles may bind to some materials or may interact with compounds eluted from the test material or other components of the test apparatus.[6] Such confounding factors may remove or inactivate virus particles so that misleading or uninterpretable data result. Thus, there must be adequate control tests to ensure that the virus is compatible with the test material, apparatus, and procedure.

2. Virus Penetration Through Used and Intact Gloves — A Review

Penetration rates have been determined in latex and vinyl gloves using HIV, HBV, HSV, as well as model viruses, such as bacteriophages. These experiments show that virus titer may be an important factor in detecting penetration,

TABLE 4   Size and Composition of Some Important Families of Human Pathogenic Viruses and Selected Surrogate Bacterial Viruses

Family	Example	Average diameter (nm)	Composition		
			Nucleic acid	Protein	Lipid envelope
Human					
Picornaviridae	Polio	30	ssRNA	Yes	No
Hepadnaviridae	Hepatitis B	42	dsDNA	Yes	Yes
Papovaviridae	Papilloma	55	dsDNA	Yes	No
Retroviridae	HIV	90–110	ssRNA	Yes	Yes
Herpesviridae	Herpes simplex	120–150	dsDNA	Yes	Yes
Bacterial					
Leviviridae	MS2	23	ssRNA	Yes	No
Microviridae	φX174	27	ssDNA	Yes	No
Pedoviridae	T7	65 (+ 17 nm tail)	dsDNA	Yes	No
Tectiviridae	PRD1	65	dsDNA (+ internal lipid mantle)	Yes	No
Cystoviridae	φ6	80	dsRNA	Yes	Yes

and extent of glove use may be an important factor in affecting the barrier effectiveness of the glove.

Klein et al.[7] tested pieces of latex, polyvinylchloride (PVC), and polyethylene exam gloves for penetration by viruses, using the bacteriophage lambda at very high titer (10^{11} plaque-forming units [pfu] per milliliter). Pieces of glove material were tested using an air sampling cassette with two chambers separated by the glove material. The virus suspension was placed in the upper chamber, and after 3 h the fluid in the lower chamber was assayed for phage penetration. At this high concentration of challenge virus, only 1 of 130 latex pieces allowed penetration of lambda virus; PVC had a high failure rate (22%), as did polyethylene (40%). At a lower concentration of challenge virus (10^8 pfu/ml), no virus penetration was found for any of the materials.

Dalgleish and Malkovsky[8] found no penetration of HIV in six brands of latex surgical gloves (two gloves tested per brand). The test procedure consisted of adding 10 ml of concentrated HIV suspension (10^4 infectious units per milliliter) to a glove, closing the glove with a knot, placing the glove inside a 60-ml syringe, and subjecting it to 20 depressions of the syringe plunger. Media was added to the syringe and then collected for assay of penetrated HIV. No HIV penetration was detected.

In contrast to the above finding, Goldstein et al.[9] found some penetration of HIV through pieces of latex that had been stretched repeatedly. Pieces of gloves were tested in a dual chamber device, with HIV on one side and medium with cells on the other. A highly polished, stainless steel rod stretched and released the latex at 60 rpm, stretching the material from two to four times its original length. Tests for penetration of HIV were done after 26, 54, 82, 117, and 156 strokes, and significant penetration was found for the last 3 numbers of strokes.

Glove tears occur during use, particularly during surgery. The overall reported failure rates for gloves in actual use by surgeons and other operating room personnel are much higher than for unused gloves. Walter and Kundsin[10] found that 20% of gloves were perforated during surgical procedures. Similarly, Church and Sanderson[11] used a water leak test to show a perforation rate of 12% following use in surgery. The perforation rates for open heart surgery were higher; Hosie et al.[12] used the same testing method as Church and Sanderson, and found an overall perforation rate of 107/252 (42%). In a 1990 study, Gerberding et al.[13] reported on perforation rates in surgical gloves used at the San Francisco General Hospital. Because this hospital is located in an area with a high prevalence of AIDS cases, surgeons almost routinely double-glove. For single-glove usage, the study found that 14 of 80 (17.5%) surgical gloves were perforated. When surgeons double-gloved, the outer glove was perforated at a rate of 78 of 448 (17.4%) gloves. The inner gloves had a perforation rate of 21 of 384 (5.5%). In another study, Wright et al.[14] report that about 11% of surgical procedures involve tears in gloves (249 glove tears in 2292 procedures). The vast majority (84%) of these tears were on the fingers. They report that 63% of the 249 tears resulted in the presence of blood on the hand.

Shombert[15] tested latex gloves that had been used by veterinarians in surgery for animals: 2 of 88 (2.3%) unused gloves and 17 of 88 (19.3%) used gloves were found to have leaks, most (14) of which were located at or within 1 cm of the tips of the fingers. Only 5 of the 17 leaks were identified by the surgeons as having leaked, as evidenced by blood on the hand after glove removal.

While the studies on glove perforation during use did not specifically address virus penetration, it is clear that blood leakage implies the possibility of penetration by free viruses and by viruses which are bound to cells or cellular components.

Kotilainen[16] suggested that glove failures may have contributed to three cases of herpetic whitlow (herpes simplex infection near the fingernail) in three nurses in a medical intensive care unit. They found that the brand of vinyl gloves used by the three nurses with herpetic whitlow had a 28% failure rate using a 300 ml water leak test (300 ml was the standard water leak challenge used before 1990).

In a subsequent investigation, Kotilainen et al.[17] tested 2000 latex and 2500 vinyl nonsterile examination gloves with both the 300 ml water leak test and the 1000 ml water leak test. An average of 10% (range 2.8 to 22.8%) of vinyl gloves, and an average of 6.4% (range 0.4 to 15.2%) of latex gloves failed the 300 ml water leak test. Additional leaks were found when these failed gloves were retested with 1000 ml of water.

The index fingers from watertight gloves were tested with buffer containing both HSV-1 and poliovirus type 1 (PV-1) (at 10^6 pfu/ml). Some gloves which passed the 300 ml water leak test subsequently allowed penetration of both types of viruses through the index fingers, with 6 to 8 times more PV-1 penetration than HSV-1 penetration. Although no HSV-1 penetration was detected from the fingers of the gloves which had passed the 1000 ml water leak test, a small percentage (1.4% for vinyl gloves, 1.5% for latex) still allowed penetration of PV-1. The overall better performance of the 1000 ml water leak test over the 300 ml water leak test prompted the authors to recommend that the 1000 ml volume be used as the standard for testing medical gloves.

Korniewicz et al.[18] tested vinyl and latex exam gloves for penetration of viruses (bacteriophage φX174), after the gloves had undergone use in a manner designed to simulate actual usage in a clinical care setting. The gloves (240 latex and 240 vinyl) were tested using 100 ml of virus-containing buffer (sufficient to just fill the fingers). Although there were usually no visually detectable leaks within 1 min, there was evidence of virus penetration during a subsequent 5-min submersion in buffer with 5 to 10% of the control gloves. Significantly different results were found only with the vinyl gloves, and only following manipulation at the highest level: 32% of these gloves leaked water and 63% allowed virus penetration.

Korniewicz et al.[19] further studied the leakage characteristics of vinyl and latex exam gloves, using the ASTM 1000 ml water leak test. They collected more than 5000 vinyl and 1000 latex exam gloves from "high risk clinical settings" (intensive care, surgery, AIDS units) and again found that used vinyl exam gloves leaked more frequently than used latex exam gloves.

3. Viral Penetration Through Punctures in Gloves

As previously described, puncture defects produced in latex surgeons' gloves with an acupuncture needle often (30 to 50% of the time) pass the FDA 1000 ml water leak test. This test is done with water inside the glove and air outside. In an environment with liquid on both sides, the possibility of liquid transfer, including accompanying virus penetration, may increase for small, undetectable holes. This possibility was tested[20] by challenging the punctured glove with 1000 ml buffer which contained the bacterial virus φX174 (27 nm diameter), conducting the (water-to-air) water leak test for 2 min, and then immersing the punctured finger into buffer for an additional 60 min to collect any viruses that may penetrate during the water-to-water test. (It should be noted that there may still remain some trapped air in some defects.)

Most of the punctured fingers which did not leak water also did not allow virus penetration; a few did allow low levels of virus penetration. However, the proportion of those that passed the water leak test but allowed virus penetration was not significantly (statistically) different between punctured fingers and control fingers.[20] Thus, it seems that the water leak test was essentially as sensitive as the virus penetration test for detecting the punctures.

However, 5.8% of the control fingers (averaged over 4 brands) had nonpuncture defects which passed the water leak test but allowed virus penetration. A typical amount of virus penetration in 60 min was equal to about 1 µl of challenge virus suspension. Using the Poiseuille equation (mentioned above) with typical values for latex thickness (100 µm) and hydrostatic pressure in the finger (12 to 15 in. of water), one can calculate a typical diameter for these "holes". The calculation shows that the average finger in a latex surgeons' glove which passes the water leak test, but allows virus penetration in our test, behaves over 60 min as though it contained a single 4 to 4.5 µm hole. This size hole would clearly not be detectable in any of the standard tests mentioned above. Furthermore, any human virus and most bacteria could pass through such a hole. Confirmatory data[6] have demonstrated that four surrogate viruses (range of diameters 27 to 80 nm) readily pass through such holes. While larger cells such as erythrocytes may deform in order to pass through such a hole, the flow rate would be significantly reduced.

The risk associated with such a hole is difficult to estimate. However, one can say that there will be intermittent short-term pressure on localized areas which may or may not contain the hole. After some use of most types of gloves, there will likely be perspiration inside the glove which may produce a water-to-water penetration situation. The concentration of free (not cell-bound) HIV rarely exceeds 10^3/ml in blood and much lower values in other body fluids, even in the viremia stages of the AIDS disease.[21,22] Thus, the risk of HIV infection from virus penetration through the undetected holes is remote, particularly since the infectivity of HIV is low[23] and the intact skin is still a very substantial barrier to disease transmission. Since hepatitis B virus titers may be several orders of magnitude higher in several body fluids,[24] penetration is more likely, and the intactness of the underlying skin becomes more important.

Attempts have been made to extend the above study with punctured latex gloves to gloves made of vinyl or nitrile. Defects produced in either type of glove with acupuncture needles have uniformly failed the 1000 ml water leak test. Thus, no marginal defects have yet been produced in these glove types.

Virus penetration through fingers of unpunctured nitrile gloves has been determined for fingers which passed the water leak test. Of 60 fingers, 1 allowed virus penetration: 0.07 µl, with a detection limit of 0.05 µl. This virus penetration is equivalent to passage of the challenge virus suspension through a single 2 µm hole (as calculated with the Poiseuille equation).

The above results and calculations suggest the need for more sensitive quality control tests for gloves. It should be noted, however, that the present 1000 ml water leak test has great utility as an overall indicator of glove quality; hence, its use as a quality control test. Furthermore, the amount of virus penetration through the average hole undetected by the 1000 ml water fill test was approximately tenfold lower than that through the barely detected hole, and several orders of magnitude lower than that through a typical hole.[20]

IV. MATERIAL INTEGRITY

A. Material Fatigue

During use, gloves are repeatedly flexed through the motions of the fingers and hands. Colligan and Horstman[25] recognized this and studied the permeation of chemotherapeutic drugs under static and flexed conditions. They found the effect of flexing on permeation depended strongly on the quality of the glove; surgical gloves were much more resistant than exam gloves.

Our experiment[31] was designed to provide very early detection of catastrophic structural changes in glove barriers rather than changes in steady-state permeability. Such changes would be caused by fatigue in the glove material due to repeated flexing. We used a modified Franz diffusion cell in which a circular section cut from a glove formed a membrane separating two chambers filled with electrolyte solutions. The membrane was repeatedly flexed at 0.15 Hz by applying varying air pressure to the top chamber. The area ratio of expanded to nonexpanded membrane was about 1.5. Assuming isotropic deformation, the linear strain was then about 22%. Introduction of electrodes into the chambers allowed the membrane capacitance and conductance to be measured continuously during the flex (fatigue) cycle. Addition of sodium ion to the upper chamber and a sodium ion electrode to the lower chamber allowed the sodium permeability to be monitored. Following detection of a catastrophic change, the membrane was examined by optical and electron microscopy using specially prepared mounts to determine the nature of the material defect which produced the change.

After about 1000 cycles (about 1.5 h) in a typical experiment, the electrical properties of the glove membrane showed significant changes. The first sign that glove failure is occurring appears in the conductance waveform. This change is shortly followed by a decrease in the membrane capacitance, showing that the latex membrane is not holding a charge; i.e., it is acting as a "leaky" capacitor. Simultaneous monitoring of the sodium ion concentration indicated a significant change in concentration over a very short timespan (1 to 3 min) relative to the total fatigue time of about 1.5 h. This change in concentration results from the diffusion of sodium ions through saline filled pores or through "thin" spots in intact glove material. However, because of the differences in the diffusion rate between intact or "thin" glove material (estimated from the change in capacitance early in the fatigue cycle) and material at final failure, it is unlikely that diffusion is through "thin spots", but through true pores instead.

Confirmatory evidence for the structural nature of the pore was provided by scanning electron microscopy (SEM). A custom mount permitted examination of the glove material in an expanded state, a state nearly identical to that in the diffusion cell. The expected agglomeration of the latex particles was readily apparent, even in the nonexpanded state. Upon expansion, however, the

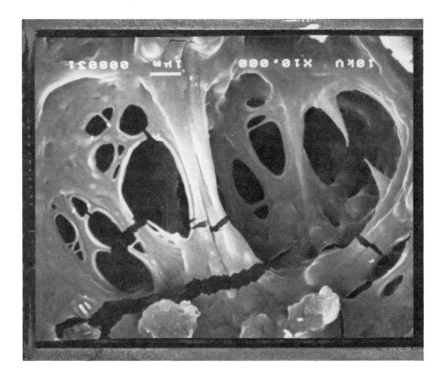

FIGURE 2 Classic "necking" of the fibril strands can be observed in electron micrograph of fatigued latex. Details of fatiguing process described in text.

nature of the connecting fibular structure becomes visible. In nonfatigued glove material, the fibrils are intact; whereas in the fatigued glove material, fibril breakage and classic "necking" of the fibril strands can be observed (Figure 2). Thus, the rapid onset of glove rupture can be explained by a classic fatigue mechanism where the glove material behaves normally until a sufficient number of fatigue cycles (and stress) lead to catastrophic rupture and hole formation. In the expanded state, the pores may approach 10,000 nm (10 μm), a size that could readily allow penetration of viruses.

B. Other Factors That Influence Integrity

The integrity of latex can be compromised by several other agents or conditions of use. Latex condoms (and by implication, latex gloves) can be affected by storage conditions, i.e., exposure to ozone, humidity, heat, or light. Studies in Indonesia[26] showed that latex condoms, packaged in plastic wrappers, and "stored under tropic conditions for approximately 42 months", performed appreciably less well than nonaged condoms in the International Standards Organization (ISO) air burst test. White et al.[27] have reported that oil-based lubricants (and, possibly, hand creams) deteriorate latex. Ozone can deteriorate latex,[28] and precautions (antioxidants) are needed during the sterilization of surgical gloves by ^{60}Co irradiation. On the other hand, storage of latex surgeons' gloves at very cold temperatures (−20°C) had no detrimental effect on the gloves' integrity or strength.[32]

IV. SUMMARY

Tests to evaluate the barrier integrity of gloves fall into two categories: those tests intended to assure quality during and after manufacture and those tests which more thoroughly challenge the barrier with viral or chemical agents but which are impractical for large quantities of gloves. The ability of all of these tests to detect microscopic holes is limited. The quality assurance tests cannot detect defects 100 times larger than HIV. Nevertheless, the viral challenges to gloves would indicate that latex provides significant barrier effectiveness to very small viruses, such as φX174. The conclusion may be drawn that, while apparent barrier integrity cannot assure safety, current quality control protocols assure that medical gloves provide significant protection.

REFERENCES

1. Douglas, A. A., Neufeld, P. D., and Wong, R. K. W., An interlaboratory comparison of standard test methods for medical gloves, in *Performance of Protective Clothing: Fourth Volume, ASTM STP 1133*, McBriarty, J. P. and Henry, N. W., III, Eds., American Society for Testing and Materials, Philadelphia, 1992, 99.
2. Carey, R., Herman, W., Herman, B., Krop, B., and Casamento, J., A laboratory evaluation of standard leakage tests for surgical and examination gloves, *J. Clin. Eng.*, 14, 133, 1989.
3. Retta, S. M., Herman, W. A., Rinaldi, J. E., Carey, R. F., Herman, B. A., and Athey, T. W., Test method for evaluating the permeability of intact prophylactics to viral-size microspheres under simulated physiologic conditions, *Sex. Transmit. Diseases* 18, 111, 1991.
4. He, L.-F., Alling, D., Popkin, T., Shapiro, M., Alter, H. J., and Purcell, R. H., Determining the size of non-A, non-B hepatitis virus by filtration, *J. Infect. Dis.*, 156, 636, 1987.
5. Lytle, C. D., Tondreau, S. C., Truscott, W., Budacz, A. P., Kuester, R. K., Venegas, L., Schmukler, R. E., and Cyr, W. H., Filtration sizes of human immunodeficiency virus type 1 and surrogate viruses used to test barrier materials, *Appl. Environ. Microbiol.*, 58, 747, 1992.
6. Lytle, C. D., Truscott, W., Budacz, A. P., Venegas, L., Routson, L. B., and Cyr, W. H., Important factors for testing barrier materials with surrogate viruses, *Appl. Environ. Microbiol.*, 57, 2549, 1991.
7. Klein, R. C., Party, E., and Gershey, E. L., Virus penetration of examination gloves, *Biotechniques*, 9, 196, 1990.
8. Dalgleish, A. G. and Malkovsky, M., Surgical gloves as a mechanical barrier against human immunodeficiency viruses, *Br. J. Surg.*, 75, 171, 1988.

9. Goldstein, A. S., Stokes, K., Fleming, W., and Arya, S. C., Small Particle Permeability of Stressed Latex Rubber Barriers, presented at Conf. Latex as a Barrier Material, College Park, MD, April 6 to 7, 1989.
10. Walter, C. W. and Kundsin, R. B., The bacteriological study of surgical gloves from 250 operations, *Surg. Gyn. Obst.*, 124, 949, 1969.
11. Church, J. and Sanderson, P., Surgical glove punctures, *J. Hosp. Infect.*, 1, 84, 1980.
12. Hosie, K. B., Dunning, J. J., Bailey, J. S., and Firmin, R. K., Glove perforation during sternotomy closure, *Lancet,* 1500, 1988.
13. Gerberding, J. L., Littell, C., Tarkington, A., Brown, A., and Schecter, W. P., Risk of exposure of surgical personnel to patients blood during surgery at San Francisco General Hospital, *N. Engl. J. Med.*, 322, 1788, 1990.
14. Wright, J. G., McGeer, A. J., Chyatte, D., and Ransohoff, D. F., Mechanisms of glove tears and sharp injuries among surgical personnel, *J. Am. Med. Assoc.,* 266, 1668, 1991.
15. Shombert, D. G., Surgical glove failure. A report on procedures similar to human surgery, *AIDS Patient Care,* December, 36, 1990.
16. Kotilainen, H. R., Latex and Vinyl Nonsterile Examination Gloves: Status Report on the Laboratory Evaluation of Defects By Physical and Biological Methods, presented at Conf. Latex as a Barrier Material, College Park, MD, April 6 to 7, 1989.
17. Kotilainen, H. R., Brinker, J. P., Avato, J. L., and Gantz, N. M., Evaluation of latex and vinyl examination gloves. Quality control procedures and implications for healthcare workers, *Arch. Intern. Med.*, 149, 2749, 1989.
18. Korniewicz, D. M., Laughon, B. E., Cyr, W. H., Lytle, C. D., and Larson, E., Leakage of virus through used vinyl and latex examination gloves, *J. Clin. Microbiol.*, 28, 787, 1990.
19. Korniewicz, D., Kirwin, M., Cresci, K., Markut, C., and Larson, E., Use of Gloves in High Risk and Low Risk Clinical Settings, Abstr. C/32, 3rd Int. Conf. Nosocomial Infections, Atlanta, GA, July 31 to August 3, 1990.
20. Kotilainen, H. R., Cyr, W. H., Truscott, W., Gantz, N. M., Routson, L. B., and Lytle, C. D., Ability of 1000 mL water leak test for medical gloves to detect gloves with potential for virus penetration, in *Performance of Protective Clothing: Fourth Volume, ASTM STP 1133,* McBriarty, J. P. and Henry, N. W., III, Eds., American Society for Testing and Materials, Philadelphia, 1992, 38.
21. Coombs, R. W., Collier, A. C., Allain, J. P., Nikora, B., Leuther, M., Gjerset, G. F., and Corey, L., Plasma viremia in human immunodeficiency virus infection, *N. Engl. J. Med.*, 321, 1626, 1989.
22. Ho, D. D., Moudgil, T., and Alam, M., Quantitation of human immunodeficiency virus type I in the blood of infected persons, *N. Engl. J. Med.,* 321, 1621, 1989.
23. Levy, J. A., The transmission of AIDS. The case of the infected cell, *J. Am. Med. Assoc.*, 259, 3037, 1988.
24. Jenison, S. A., Lemon, S. M., Baker, L. N., and Newbold, J. E., Quantitative analysis of hepatitis B virus DNA in saliva and semen of chronically infected homosexual men, *J. Infect. Dis.*, 156, 299, 1987.

25. Colligan, S. A. and Horstman, S. A., Permeation of cancer chemotherapeutic drugs through glove materials under static and flexed conditions, *Appl. Occup. Environ. Hyg.*, 5, 848, 1990.
26. Free, M. J., Hutchings, J., Lubis, F., and Natakusumah, R., An assessment of burst strength distribution data for monitoring quality of condom stocks in developing countries, *Contraception,* 33, 285, 1986.
27. White, N., Taylor, K., Lyszkowski, A., Tullett, J., and Morris, C., Dangers of lubricants used with condoms, *Nature,* 335, 19, 1988.
28. Baker, R. F., Sherwin, R. P., Bernstein, G. S., Nakamura, R. M., Voeller, B., and Coulson, A., Precautions when lightning strikes during the monsoon. The effect of ozone on condoms, *J. Am. Med. Assoc.,* 260, 1404, 1988.
29. Herman, B. A. and Carey, R., unpublished data, 1991.
30. Retta, S. M. and Rinaldi, J. E., unpublished data, 1992.
31. Dillon, J. G. and Schroeder, L. W., unpublished data, 1993.
32. Lytle, C. D. and Marlowe, D. E., unpublished data, 1992.

PART IV

CLINICAL TESTS WITH PROTECTIVE GLOVES

10

Diagnosis-Driven Management of Natural Rubber Latex Glove Sensitivity

Curtis P. Hamann and Shelley A. Kick

TABLE OF CONTENTS

I. Introduction 131
II. Types of Reactions 132
    A. Irritant Reactions 132
    B. Delayed Allergic Reactions 133
    C. Immediate Allergic Reactions 134
III. Diagnostic Evaluation 135
    A. Risk Factors 135
    B. Physical Examination 136
    C. Allergy Tests 137
        1. Patch Testing 138
        2. Skin Prick Test 139
        3. Use Test 142
        4. Intradermal Testing 142
        5. Radioallergosorbent Test (RAST) 143
    D. Differential Diagnosis 145
IV. Management 147
V. Conclusion 149
References 152

I. INTRODUCTION

Gloves protect workers from many occupational hazards. In health care, where gloves form an integral part of universal precautions, cross-contamination, viral penetration, and permeability to antineoplastic drugs and infection control agents are the primary concerns. In occupations as disparate as industry and cosmetology, gloves protect workers from a variety of chemical hazards. In the food industry, infection control again becomes the major focus. Domestic gloves from natural rubber latex (NRL) may be used in the home or by sanitary workers. The livelihood of many of these workers depends on their ability to wear gloves. What, then, are the options for workers when their protective gloves become an occupational hazard?

In most cases, NRL gloves are the culprits. Dermatitis from NRL gloves — attributable to chemical additives — has been known for at least 60 years.[1] Since Nutter's[2] first report of contact urticaria to NRL gloves in 1979, overwhelming evidence has indicated that proteins associated with the latex can induce hypersensitivity. Despite the increasing reports of allergic reactions to NRL gloves, confusion still exists about the causes, the diagnostic evaluation, and the treatment of such sensitivities. A recent survey suggests that the specific allergen underlying most reactions to gloves is almost never diagnosed by a professional.[3] Clinicians, however, must understand these issues to insure that patients receive appropriate care and to prevent future sensitization. This chapter, therefore, reviews the symptomatology, diagnostic evaluation, and management of patients who present with NRL glove-related sensitivities.

II. TYPES OF REACTIONS

The clinical manifestations of glove sensitivities reflect three distinct mechanisms: irritation, delayed reactions, and immediate allergic reactions. Diagnosis — and hence, the appropriate treatment — can be complicated by the overlapping clinical picture of these three reactions. Formulating the optimum treatment strategy for patients requires understanding the types of reactions and the irritants and allergens in gloves. Hypersensitivity can be induced by allergens that are a natural component of NRL, by allergens that are the product of the manufacturing process, or by a combination of the two.

A. Irritant Reactions

An irritant reaction is a nonallergic response to contact with an external agent that damages dermal cells (see Chapter 15).[4] Irritant contact dermatitis, the characteristic inflammatory response of the skin, can be induced by an excess of processing chemicals such as accelerators, surfactants, or preservatives (see below).[5] Excess chemicals can result from insufficient leaching or curing of gloves. In turn, leaching is a function of temperature, duration, ingredients, and enzyme treatment. Proper curing depends on temperature, duration, and the mix of accelerators and vulcanizing agents.

Other features of gloves can also cause irritant contact dermatitis. The glove powder used as a donning lubricant can macerate the occluded hand and cause mechanical irritation.[6,7] Irritant reactions can also be the result of procedures used to sterilize gloves. Gamma irradiation of gloves releases bacterial endotoxins, which have been implicated as a factor in contact dermatitis in at

TABLE 1 Agents Added to Natural Rubber Latex Gloves During Manufacturing

Accelerators	Germicides
Antiblocking agents	Gelling agents
Antifoaming agents	Wetting agents
Antiwebbing agents	Homogenizing agents
Antiflex cracking agents	Inhibitors
Antioxidants	Lubricants
Antistaining agents	Modifiers
Chain extenders	Odorants
Optical brighteners	Plasticizers
Chemical and heat sterilizers	Coagulants
Softeners	Dispersants
Vulcanizing agents	Preservatives
Dusting materials	Recleaning materials
Dipping materials	Reducing agents
Washing materials	Retardants
Emulsifiers	Solvents
Finishers	Terminators
Flame retardants	Ultraviolet inhibitors
Fungicides	

least one report.[8] Ethylene oxide, also used to sterilize gloves, is another potential irritant.[9]

B. Delayed Allergic Reactions

Clinically and pathologically, irritant dermatitis is not easily distinguished from allergic contact dermatitis. Both share a similar constellation of symptoms: redness, swelling, papules, perivascular infiltration of lymphocytes, and edema.[4] The mechanisms underlying the two reactions, however, are quite different. Unlike irritant contact dermatitis, allergic contact dermatitis reflects an acquired immune reaction mediated by T cells and macrophages.[10,11] This response is also known as type IV hypersensitivity (see Chapter 16).

Delayed allergic reactions are primarily the result of agents added to latex during the process of manufacturing.[11] As many as 200 different chemicals can be added to the formulation of a NRL-dipped product such as a glove (Table 1).[12] These chemical additives can account for 5% of the weight of a finished product. Although any of these agents has the potential to cause contact dermatitis, only a few groups actually appear to induce skin reactions.

Two major classes of chemicals are responsible for dermatological problems in glove wearers — accelerators and antioxidants.[12] Accelerators are added to insure uniform and relatively rapid vulcanization of NRL. Antioxidants are added to delay the inevitable deterioration of NRL, which is caused by common factors found in the environment (e.g., ozone, heat, humidity) as well as by mechanical stress.[13] Many of the additives decompose at various stages of the manufacturing process and pose no risk to the user. Some antioxidants and accelerators, however, persist in the final product. These chemicals can then "bloom" onto the surface of the glove where they contact the hand,[14] creating the opportunity for sensitization. Thiurams, carbamates, thiazoles, and thioureas, the most common groups of chemical additives, are present in unique ratios and concentrations in different brands of gloves. These variations are proprietary formulations that reflect manufacturing compromises needed to obtain the desired level of such qualities as tensile strength, elasticity, durability, and tactility.

Historically, thiurams have been responsible for most contact reactions. For example, of 356 patients with glove dermatitis studied between 1971 and 1976, 90% were sensitized to thiurams.[15] In a more recent study by Hintzenstern et al.,[16] thiurams accounted for 62% of the occupationally related glove allergies. Consequently, many manufacturers of latex surgical and examination gloves have eliminated thiurams from their products by substituting less sensitizing accelerators.

Carbamates are now the most common accelerators found in NRL gloves and the second most common cause of contact dermatitis.[6] In the Hintzenstern et al. study,[16] carba mix was responsible for 25% of the occupationally related glove allergies. Carbamates alone may induce delayed sensitivity in individuals. However, they are chemically related to thiurams, with which they exhibit cross reactivity.[17] Carbamates also induce irritant reactions that can be difficult to distinguish from a delayed allergic reaction. During the manufacture of gloves, special washing procedures are used to maintain carbamate concentrations below 0.5% to minimize these problems.[6]

The third most common allergen in NRL gloves is mercaptobenzothiazole. In the Hintzenstern et al. study,[16] mercapto-mix accounted for 3% of the glove reactions. Other members of the thiazole group of accelerators are restricted to products like tires and shoes.[14] Thioureas, another group of accelerators and potential allergens, appear to be used in very few gloves.[6]

C. Immediate Allergic Reactions

NRL gloves can also induce immediate type I hypersensitivity,[18,19] an IgE antibody-mediated response.[4] Naturally occurring proteins in latex, rather than chemicals added during processing, are the primary source of type I reactions (see Chapter 17). NRL is the intracellular fluid produced by the laticifer cells of the rubber tree, *Hevea brasiliensis*.[20] In addition to the functional rubber molecule, *cis*-1,4 polyisoprene, NRL contains carbohydrates, lipids, phospholipids, and proteins. The total protein content of NRL is about 1.7%, 60% of which is associated with the polyisoprene.[21]

Three proteins in NRL have been identified. At 38 kDa, prenyltransferase, a water-soluble protein, is the largest. Together with a smaller protein, rubber elongation factor (14.5 kDa), prenyltransferase helps synthesize polyisoprene.[22] The third protein, hevein, is small (10 kDa) and crystalline.[23] Despite the identification of these three proteins, the identity of the latex allergen is unclear. Variability in antigen source (ammoniated and nonammoniated NRL as well as extracts from a variety of finished products), isolation technique [radioallergosorbent test (RAST), immunoblot, basophil histamine release, high pressure liquid chromatography (HPLC)], and serum pools have precluded definitive characterization of the latex allergen,[24-33] but proteins from 2 to 200 kDa have been reported. The proteins most likely to have clinical relevance are concentrated between 14.5 and 30 kDa.

Although the dermatologist will typically be confronted with patients exhibiting either contact dermatitis or contact urticaria, the potential for a more serious reaction should not be underestimated. At least 15 patients have died from anaphylactic reactions induced by the rubber catheter tips used to perform barium enemas.[34-36] Such serious reactions are unlikely during skin testing if proper technique is used, but the potential should be recognized. Patients have been reported to have experienced anaphylaxis after exposure to NRL in normally innocuous settings such as skin testing.[37]

III. DIAGNOSTIC EVALUATION

The diagnosis of glove-related sensitivities depends on taking a careful history, physical examination, and allergy testing. The history should seek to establish the presence of key risk factors, which when combined with the findings from the physical examination, will dictate the need for further testing. The choice of a provocative or *in vitro* test will, in turn, reflect the patient's risk for a serious reaction as indicated by the history. A careful history cannot be expected to identify all patients with the potential to react to NRL or to eliminate the need for confirmatory testing. Kwittken and Sweinberg,[38] however, found that half of their pediatric patients had a clear history of relevant risk factors. Patients presenting with contact dermatitis or contact urticaria should therefore be questioned closely about the following risks.

A. Risk Factors

The risk factors for NRL allergies are well defined and represent both environmental exposure and genetic predisposition. The groups at highest risk for delayed allergic reactions are workers in the health care industry (e.g., nurses, hospital personnel, dentists), rubber industry, or other occupations where gloves are worn (e.g., food preparation, cosmetology). For example, the U.S. Food and Drug Administration (FDA) estimates that 6 to 7% of hospital surgical personnel are sensitive to NRL.[39] In a Finnish teaching hospital, 7.4% of surgeons and 5.6% of surgical nurses had allergic reactions to NRL gloves.[40] In a Canadian glove factory, 8.6% of the workers experienced contact urticaria on the job[41] although the reproducibility of this result in other factories is controversial. In a recent survey of the U.S. Army Dental Corps,[3] 13.7% of the respondents reported signs and symptoms of NRL glove allergies. Even if the nonrespondents of this study were considered allergy free, the prevalence of symptoms would still have been 8.8%. These groups, which are also at risk for immediate allergic reactions, are most likely to seek treatment for glove-related reactions.

In general, children who have undergone multiple surgeries during infancy or childhood have the highest risk of developing an immediate allergic reaction to NRL. The reported incidence is by far the highest (18 to 40%) among children with spina bifida.[33,42,43] Other high-risk groups include patients with congenital urogenital abnormalities, spinal cord injuries, or conditions that require multiple reconstructive surgeries.[44] Suspicion of an allergy to NRL should be increased if patients report an intraoperative anaphylactic reaction of undetermined cause or any previous reaction to NRL.

The prevalence of NRL sensitivity is higher among females than males.[45] Individuals with atopy, especially those with hand eczema, also appear to be at risk for developing sensitivity to NRL.[18,46-48] Some atopics with NRL allergies have reported a history of fruit allergies. These cases suggest a cross-reactivity between NRL proteins and bananas,[49-51] avocados,[52,53] chestnuts,[49] and passionfruit.[54] None of these plants share a close taxonomic relationship with *H. brasiliensis,* and it has not yet been established if the presence of these food allergies is predictive of NRL sensitivity.

Among these disparate risk groups, the common feature appears to be the level of exposure, whether occupational or therapeutic. Because NRL allergy is a relatively new phenomenon, other risk factors could still be identified.

B. Physical Examination

Diffuse or patchy eczema on the dorsal surface of the hands, fingers, wrists, and distal forearm,[11,12,55] combined with occupational or routine therapeutic use of gloves, should heighten the suspicion of an allergic reaction to gloves. In a classic case of glove dermatitis, the symptoms will not extend beyond the wrist, but a nonspecific pattern of hand eczema can occur. Delayed allergic contact dermatitis typically appears as redness and inflammation at the site of exposure 48 to 96 h later. Vesicles and blisters may also form. Such findings suggest the need for routine patch testing. Typically, these patients will have a delayed eczematous glove dermatitis to one or more of the chemicals added to NRL gloves. Irritant contact dermatitis can be difficult to distinguish from allergic contact dermatitis. Atopic individuals are especially susceptible to irritant reactions, which can mimic the clinical manifestations of a true allergic reaction.

A physical examination can also provide clues about contact urticaria from NRL gloves (Figure 1). In general, the clinical manifestations of an immediate allergic reaction include local and systemic urticaria, rhinitis, conjunctivitis, bronchospasm, and anaphylaxis, depending upon the site of exposure to the antigen.[37,56,57] Contact urticaria from NRL gloves may be associated with pruritis, stinging, or discomfort over the hands. Redness, swelling, or a wheal-and-flare response may begin 5 to 20 min after a patient dons a glove. Without treatment, the response typically subsides within 1 to 2 h. Associated systemic

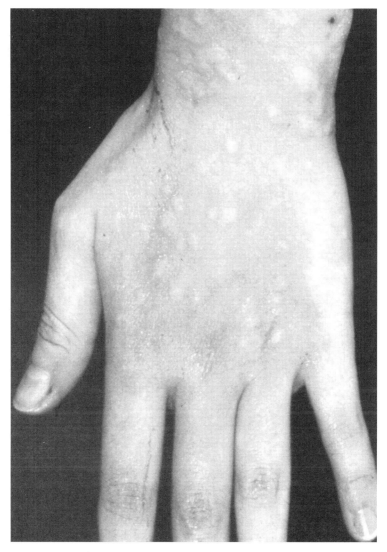

FIGURE 1 An immediate, IgE mediated contact urticaria induced by NRL gloves. (Courtesy of Kristiina Turjanmaa, M.D., Ph.D.)

symptoms such as rhinitis, dizziness, or angiodema should alert the physician to the potential for more serious reactions. Patients who exhibit either constellation of symptoms, in the presence of one or more of the risk factors, should be considered for further allergy testing.

C. Allergy Tests

In most cases, an allergy test will be required to confirm the diagnosis of a sensitivity to NRL or rubber chemicals and to enable the physician to recommend appropriate therapeutic options. Given the potential for anaphylactic shock, the physician should proceed cautiously with all *in vivo* testing.[58,59] Emergency drugs and resuscitative equipment should be readily available. If the physician suspects that the patient has previously experienced a severe systemic reaction to NRL, *in vitro* testing is a diagnostic alternative.

Both *in vivo* and *in vitro* tests have advantages and disadvantages that must be considered when interpreting the results.[60] *In vivo* testing can further sensitize patients, boost the immune response, or cause a flare-up at a previous reaction site. *In vitro* studies, which avoid the risks of interfering with the immune system, tend to be less sensitive. The sensitivity and specificity of each test vary not only with the inherent weakness of the test itself, but also with the source of the allergen used for testing and with the condition of the patient's skin. Unfortunately, no single test is 100% predictive, and the unique problems associated with each should be remembered before pursuing an individual diagnostic course. The following discussion concentrates on the tests most commonly used to diagnose reactions to rubber chemicals or NRL.

1. Patch Testing

Overall, effective *in vitro* tests for detecting delayed reactions have yet to be developed. Patch testing is therefore the standard diagnostic procedure for type IV reactions. The general indications for a patch test are dermatitis that persists for more than a month or a reaction that interferes with a worker's ability to perform his or her job.[61]

The test units for the patch test are either 10-mm circular filter paper disks with aluminum backs or Finn Chambers® — flat aluminum cups, 10-mm in diameter.[61] The test substance is usually dissolved in one of three vehicles: petrolatum, water, or ethanol. In the case of rubber chemicals, the preferred vehicle is petrolatum. The unit should be applied with Dermacel tape to avoid interfering with the reading. The preferred test sites are the back and the ventrolateral sides of the arm. Patch tests are typically read within 24 to 48 h. A positive reaction — itching and papulovesicles that persist for one or two days to a week — suggests that the patient is sensitive to the test allergen. A reaction that persists more than 6 days may be considered a flare-up.

Although patch testing is the diagnostic standard, it can produce misleading results.[61] Exposure to the test substance carries a small but definite risk of sensitization. If the concentration is too high, an irritant reaction may occur. If the allergen concentration is too low, the risk of a false negative is increased. Therefore, negative responses cannot be considered diagnostic. False negatives can also occur if the allergen fails to penetrate the skin because the occlusion is inadequate or too little material is applied to the patch. An inappropriate vehicle is another cause of false negatives. A standard battery for rubber chemicals eliminates the uncertainty about choice of vehicle or allergen concentration.

The standard tray for rubber chemicals consists of four mixes of the most common accelerators and antioxidants in petrolatum (Table 2).[14] The concen-

TABLE 2 Rubber Mixes Used in Standard Tray

Chemical	Petrolatum (%) NACDG	Petrolatum (%) ICDRG
Thiurams (thiuram mix)		
Tetramethylthiuram monosulfide (TMTM)		0.25
Tetramethylthiuram disulfide (TMTD)		0.25
Tetraethylthiuram disulfide (TETD)		0.25
Dipentamethylenethiuram disulfide (PTD)		0.25
Thiazoles (mercapto mix)		
2-Mercaptobenzothiazole (MBT)	1	0[a]
	—	0.5[b]
N-cyclohexyl-2-benzothiazyl sulfenamide (CBS)	0.33	0.5
Dibenzothiazyl disulfide (MBTS)	0.33	0.5
Morpholinyl mercaptobenzothiazole (MOR)	0.33	0.5
Diamines (PPD mix)		
N-isopropyl-N'-phenyl-p-phenylenediamine (IPPD)	0.1	
N,N'-diphenyl-p-phenylenediamine (DPPD)	0.25	
N-phenyl-N'-cyclohexyl-p-phenylenediamine (CPPD)	0.25	
Dithiocarbamates (carba mix)		
1,3 Diphenylguanidine (DPG)[c]	1	
Zinc dibutyldithiocarbamate (ZDBC)	1	
Zinc diethyldithiocarbamate (ZDEC)	1	

Note: NACDG = North American Contact Dermatitis Group, ICDRG = International Contact Dermatitis Research Group.

[a] Tested separately in NACDG tray.
[b] Tested separately and in mix in ICDRG tray.
[c] Included in carba mix although not a carbamate.

From Brandão, F. M., *Occupational Skin Disease,* 2nd ed., Adams, R. M., Ed., W. B. Saunders, Philadelphia, 1990, 462. With permission.

trations of the individual chemicals in each mix are intended to be below the threshold for irritation; consequently, false negatives do occur. The thiuram and diamine mixes are reliable detectors of sensitivity. Carba mix occasionally produces a weak irritant reaction. Some patients have tested positive to mercaptobenzothiazole at 1%,[62] but negative to the mercapto mix. Consequently, the North American Contact Dermatitis Group removed this chemical from the mix and recommends testing for it separately.[14] In Europe, however, mercaptobenzothiazole is maintained in the mix.

2. Skin Prick Test

In the skin prick test, antigen diluted in saline is dropped on the skin and gently pricked with a lancet (Figure 2). The development of a wheal-and-flare response is observed across time and compared with a control group reaction. The appearance of a wheal-and-flare within 10 to 60 min suggests a type I reaction. A 4- to 6-h delay in the onset of the response suggests a type IV reaction.[60]

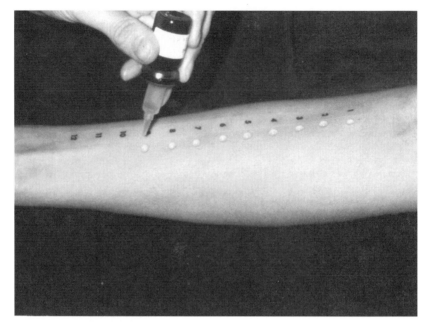

A

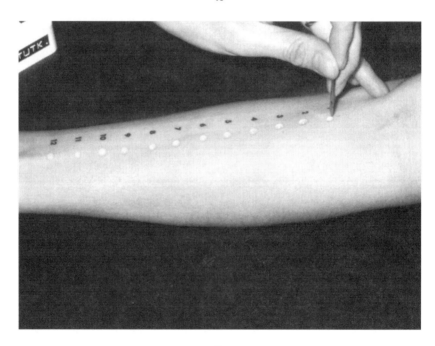

B

FIGURE 2 (A) In the skin prick test, antigen diluted in saline is dropped on the skin, and (B) gently pricked with a lancet. (C) Positive reaction. (Courtesy of Kristiina Turjanmaa, M.D., Ph.D.)

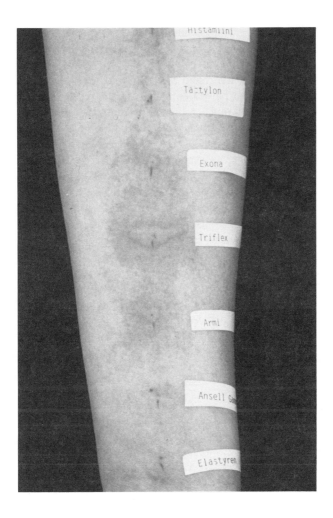

C

The skin prick test is easy to perform and readily available. It is highly sensitive and provides quick results. With NRL, however, the test also has the potential for inducing a full systemic reaction. This disadvantage partially reflects the provocative nature of the test and partially reflects the absence of a standard NRL allergen. To minimize the opportunity for an adverse reaction, Sussman et al.[59] recommend wiping the skin clean with isopropyl alcohol immediately after a positive result, which could be as soon as 5 to 7 min after exposure. Currently, physicians must prepare their own NRL extracts, and an additional element of risk is thereby introduced. The relationship between sensitization and the amount of protein in a given product is unknown, as is the concentration and availability of protein for extraction in different products.[27] Furthermore, variations in the techniques used for extraction can affect the final allergen concentration unpredictably. Greater standardization is clearly needed.

Turjanmaa and Reunala[63] have used the following extraction technique successfully: 20 squares (1 cm × 1 cm) cut from NRL gloves are incubated in 5 ml of sterile saline for 30 min at room temperature, and the resulting solution is used as the test allergen; histamine dihydrochloride (10 mg/ml) and saline serve as positive and negative controls. The volar side of the arm is pricked. After 15 min, the wheal is measured in the two largest perpendicular axes. These measurements are summed and divided by two. A positive glove reaction is greater than or equal to half of the value of the histamine control.[12]

Anaphylactic reactions to skin prick tests have been reported[64] and underscore the need for close observation of the patient during the testing. Differences in the severity of reactions may reflect the variability in the allergenicity of the eluates derived from different glove products[63] — a possibility that has received little systematic investigation. However, if the reagent is prepared properly and the technique executed correctly, the skin prick test is the most convenient and sensitive diagnostic method. In the U.S. the FDA is urging manufacturers to develop a standard NRL antigen.[65]

3. Use Test

In this provocative test, the patient dons one finger of a glove (Figure 3).[61] Itching or urticaria within 30 min indicates a positive response. If there is no reaction, the patient dons an entire glove on a wet hand for another 30 min. During the challenge of both the finger and the entire hand, the patient can wear a control glove of a synthetic material (e.g., vinyl, Tactylon™) to rule out nonspecific erythema or dermographism. Prudence dictates testing the finger before the entire hand is challenged because severe urticarial[54] and anaphylactic reactions[27] from the initial application of a glove have been reported. The most conservative approach would therefore be to reserve the use test to confirm a weak positive prick test.

4. Intradermal Testing

Intradermal testing involves injecting a very small, very dilute amount of the antigen into the patient's lower arm.[66] Again, no standard NRL allergen is available. Localized erythema after 10 min is considered a positive response. In general, intradermal testing may be 1000 times more sensitive than prick testing. Unfortunately, anaphylactic responses are also more common with intradermal testing than with prick testing.[67] Given the controversy surrounding the safety of using the prick test to diagnose NRL sensitivity, intradermal testing is difficult to justify. Intradermal testing has recently been suggested for the preoperative assessment of high-risk surgical patients with suspected IgE-

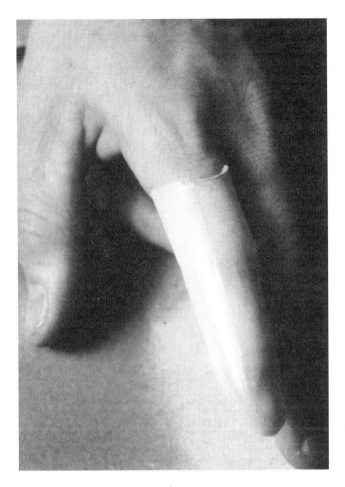

A

FIGURE 3 (A) During the use test, the patient dons one finger of the glove suspected of inducing reaction. (B) If there is no reaction within 30 min, the patient dons an entire glove while wearing a synthetic control glove on the other hand. (C) Positive reaction on finger. (Courtesy of Kristiina Turjanmaa, M.D., Ph.D.)

mediated responses to NRL.[44] This test, however, cannot be recommended as part of the routine diagnostic evaluation of glove sensitivities.[59]

5. Radioallergosorbent Test (RAST)

The RAST is a commercially available (Pharmacia, Piscataway, NJ) *in vitro* test that measures latex-specific IgE antibodies.[68] The allergen is incubated with the patient's serum to induce specific antigen-antibody reactions. ^{125}I-labeled anti-IgE is then added to the serum to detect the allergen-specific antibodies. RAST is the most specific test for IgE antibody;[37] however, its

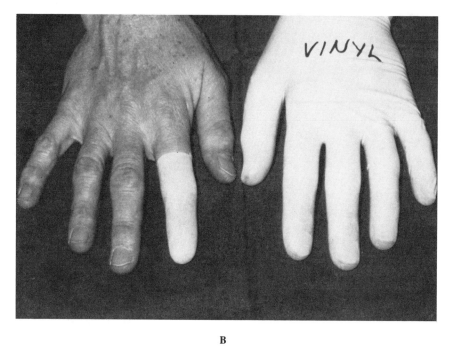

B

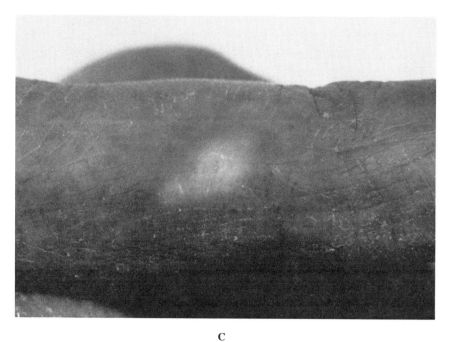

C

FIGURE 3 (continued).

sensitivity, which ranges from 40 to 70%, limits its clinical utility.[29,69] Its greatest applicability is probably in confirming type I reactions, but a negative *in vitro* test cannot rule out sensitivity to a specific allergy. Because of its specificity in detecting highly allergic patients, RAST can be used to confirm the diagnosis of NRL sensitivity in patients suspected of hypersensitivity based on a history of anaphylaxis.[70] Theoretically, RAST could be used as a clue for further topical testing of antigens, but its expense and restricted availability limit its practicality as a noninvasive screening tool.

Other *in vitro* tests have been used to investigate the antigens associated with gloves: the enzyme-linked immunosorbent assay (ELISA),[71-73] basophil histamine release,[71,72,74] lymphocyte proliferation,[73] immunoblot,[30] radioimmunoassay,[18] cross and rocket immunoelectrophoresis,[24] and reverse enzyme immunoassay.[74] Currently, these tests have little clinical relevance. The reader interested in the role of these tests in research should refer to Chapter 17.

D. Differential Diagnosis

If an occupational dermatitis is suspected, other sources of allergens in the workplace, besides gloves, must also be considered. For example, food hypersensitivity cannot be discounted among workers in the food industry. Among health care workers, frequent handwashing is unavoidable. Exposure to a variety of detergents and disinfectants can also cause allergic contact dermatitis[55] that could be difficult to differentiate from glove-related reactions. If a patient works in a medical or dental field, patch testing may be necessary to differentiate allergic reactions to common antiseptic agents from glove dermatitis or urticaria (Table 3).

Other chemicals are another potential source of occupational contact dermatitis. For example, orthopedic surgeons or dentists can develop an allergic contact dermatitis to an unpolymerized residue of the methyl methacrylate monomer used to mix acrylic bone cement.[55] If mixed or applied by hand, the cement readily penetrates NRL gloves and can damage nerve fibers. The concomitant presence of paresthesia in the fingers is therefore a distinguishing symptom. Sensitized patients will also show a strong positive patch test response to methyl methacrylate. Formaldehyde, another strong cutaneous irritant widely used in hospitals and laboratories, has also been reported to cause allergic contact dermatitis.[55,75] Antineoplastic drugs are a potential source of irritant contact dermatitis among hospital workers.[55] The possibility of occupational exposure to an alternative allergen underscores the importance of obtaining a thorough history of the patient.

A common misconception is that glove powder is the offending allergen. Health care workers themselves frequently make this assumption.[40] The powder in gloves is a cornstarch derivative added to help release the gloves from the mold during manufacturing as well as to help users don the gloves. Pure cornstarch powder appears to be an extremely rare sensitizer, although it can

TABLE 3 Antiseptics Associated With Allergic Contact Dermatitis

Agent	Types of products	Examples of use	Clinical manifestations	Test
Glutaraldehyde	Cidex™	Disinfect endoscopes	Contact dermatitis	Patch, 1% aqueous
Chlorhexidine	Hibiclens Corrodyl Hibidil Hibistat	General antiseptic	Eczema Contact dermatitis Contact urticaria Anaphylaxis	Patch, 0.5% Prick, 0.005%
Povidone-iodine	Betadine	General antiseptic	Irritant contact dermatitis	Patch, undiluted
Alcohols	Various	Various	Eczema Erythema Contact urticaria	Patch, undiluted
Benzalkonium chloride	Various over the counter products Ophthalmic solutions	Preoperative skin disinfectant Surgical instrument disinfectant	Contact dermatitis	Patch, 0.1% aqueous
Hexachlorophene		Preoperative scrub	Contact dermatitis	Patch, 1% petrolatum

From Fisher, A. A., *J. Allergy Clin. Immunol.*, 90, 729, 1992. With permission.

cause mechanical irritation or exacerbate existing dermatoses. Western blot analysis has shown that cornstarch can, however, absorb glove allergens.[76] This new hapten, representing NRL proteins bound to starch particles, may have an increased potential to act as an allergen. These particles may become aerosolized when packages of NRL gloves are opened, thereby contaminating large work areas and increasing the overall risk of sensitization to NRL on the job. Because the allergens become airborne, sensitized individuals may often report respiratory symptoms. To rule out an allergic reaction to pure cornstarch, a small amount of the powder obtained from a nonglove source can be rubbed into intact skin on the patient's forearm. The appearance of a wheal within 20 min indicates a positive response.[55]

Although many complicated factors must be considered to diagnose a glove reaction correctly, the diagnostic process can be summarized by the algorithm presented in Figure 4. The first step is to obtain a complete history and to determine if a patient is a member of a high risk group. If no risk factors can be established, testing for NRL or rubber chemical sensitivity may not be warranted and other diagnoses should be entertained.

If the patient's history establishes a reasonable suspicion for a glove allergy, the likelihood of a type IV reaction to one of the processing chemicals should be evaluated first by performing a patch test with the standard battery of rubber chemicals. A positive response identifies a specific antigen but does not rule out a simultaneous type I reaction to NRL. The most conservative approach to diagnose a type I NRL allergy would be to first perform a RAST. This choice is desirable for any patient with unexplained severe systemic responses that might

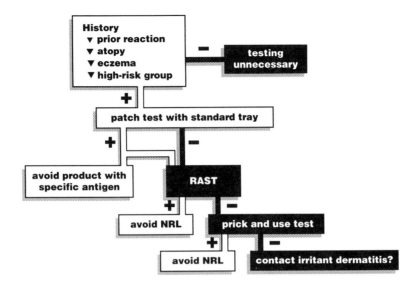

FIGURE 4 Diagnostic and treatment algorithm for type I and type IV reactions related to natural rubber latex.

be linked to NRL (e.g., severe bronchial reaction after blowing up a balloon or exposure to a dental dam). A positive RAST confirms a latex allergy. Because of its poor sensitivity, however, a negative RAST cannot be considered diagnostic. Prick and use tests are then necessary. The use test can be used to confirm a weak positive response on the prick test. If both the prick and use test are negative, the likelihood of contact irritant dermatitis is increased.

IV. MANAGEMENT

Once the allergen has been diagnosed correctly, the physician can devise an appropriate management strategy for the patient. Too often patients are undiagnosed or, if a glove reaction is suspected, simply told to change their glove. Trial and error in finding an alternative glove are substituted for avoiding a glove based on knowledge of its contents.

The consequences of such an approach can be serious. Undiagnosed or untreated allergies can increase absenteeism. In the worst case, highly skilled and trained professionals such as health care workers can be forced to abandon their careers. Employers may have to bear the burden of training replacement workers or may be encumbered with insurance, medical, or disability costs if the condition worsens. Continued exposure may increase the risk of a life-threatening anaphylactic reaction. As with any disorder, a diagnosis enables the physician to make appropriate recommendations and the patients to make informed choices that are relevant to their livelihood and well-being.

The diagnoses, however, must be accompanied by a clear understanding of the potential allergens in the various brands of gloves.[6,12] The physician must therefore help educate the patient on two important issues related to choosing a glove. The first issue is simple to conclude but more difficult to implement: patients must learn to recognize and avoid products that contain the offending allergen(s). From a broad perspective, patients must learn to assume the responsibility of making queries about the contents of a product. Unfortunately, with gloves this task is far from simple. Currently on the market are dozens of brands of gloves, the formulations of which vary from manufacturer to manufacturer. Furthermore, different manufacturers may supply gloves that are labeled as one brand — a practice that increases the uncertainty about glove constituents. No FDA regulations yet require product labeling so consumers cannot ascertain the constituents of gloves from the packaging. This information can only be obtained by contacting the manufacturer or combing the scientific literature — neither of which represents a user-friendly avenue. The partial listing provided in Table 4 can be shared with patients with glove sensitivities to assist in their search for allergen-free gloves. Patients' needs and their preferences in fit and feel may vary with their occupation; these features likewise vary from brand to brand. Ideally, patients should have several options, not only to insure allergen avoidance, but to guarantee their comfort in these other dimensions that are also important in the wearing of protective gloves.

The indications for choosing a particular glove are straightforward, but the choices narrow as the number of allergens a patient is sensitive to increases (Table 4).[6] Gloves containing carbamates can be worn by patients allergic to thioureas and thiazoles. Gloves containing carbamates could also be worn by patients allergic to thiurams, but the possibility of a cross-reaction should be considered. Gloves containing carbamates and thiazoles can be worn by patients who react to thiurams and thioureas. Gloves containing only benzothiazoles can be worn by patients with allergies to thiurams, carbamates, or thioureas. The combination of allergens to be avoided will dictate the best choice of glove, but it should be remembered that all NRL gloves contain at least one accelerator and very often two.

If NRL proteins, rather than the processing chemicals, are the allergens, patients' choices are restricted to the few available types of synthetic gloves.[12] Vinyl gloves are free of both NRL and accelerators, but their feel and fit limit their utility to nonsurgical activities. In this age of universal precautions, concern has also been expressed about the efficacy of the barrier protection offered by vinyl gloves.[77-79] Nonetheless, high quality vinyl gloves are available and may be an appropriate choice in many settings for individuals who are sensitive to both NRL and accelerators. Other alternatives for NRL-sensitive individuals are gloves with inner coatings. Regent's Biogel® is lined with hydrogel and Neutralon® is lined with polyurethane.[6] Unfortunately, inner linings can deteriorate and expose the wearer to the outer layer of NRL. They cannot therefore be recommended for patients who are hypersensitive to NRL proteins.

Several other synthetic gloves could also be used by NRL-sensitive individuals. Elastyren® and Neolon® are free of NRL but contain carbamates. Dermaprene is also free of NRL but is formulated with the accelerator diphenylthiourea. The only nonvinyl glove that is entirely free of NRL and all accelerators is Tactyl-1™. This glove may represent the only suitable alternative for health care workers with both type I and type IV sensitivities to NRL gloves. A five-layer plastic glove containing polyethylene and an ethylene-vinylalcohol-copolymer can be used by patients who are sensitive to glove-permeating chemicals.[6]

Nonpowdered gloves are available for the rare individual who exhibits an allergy or irritant reaction to glove powder. Cotton underliners are an excellent option if irritant contact dermatitis is indicated by negative allergy tests.

Patients should also be educated to regard the label, hypoallergenic, with skepticism.[7,12] "Hypoallergenic" should not be confused as meaning free of powder, NRL, or accelerators. Hypoallergenic gloves can contain any combination of these substances, albeit at reduced levels. The only requirement for making a claim of hypoallergenicity is a Modified Human Draize Test in 200 random subjects. At best this test is a gross screening tool. The results cannot be used to predict the true allergic potential of a glove, especially among previously sensitized individuals. Hypersensitive individuals could still react to the low concentrations of protein allergens found in gloves carrying the hypoallergenic label.

The FDA is currently reexamining the criteria that should be satisfied before a claim of hypoallergenicity can be made.[65] Manufacturers may be required to carry a warning label stating that medical devices contain NRL. These steps would ease the consumer's dilemma in choosing a glove. Even under these circumstances, however, information on the level of the various allergens is unlikely to be provided. The physician as well as the patient should be aware of the potential danger in prescribing a glove on the basis of a claim of hypoallergenicity alone.

V. CONCLUSION

The diagnosis of type IV sensitivities to rubber chemicals has the benefit of a long history and is relatively straightforward. Nonetheless, continued investigation of glove additives with allergenic or irritant potential could clarify the process of diagnostic evaluation when the results of skin tests are equivocal or negative. Many questions, however, remain to be answered before the testing and treatment of NRL reactions can be considered definitive. Because universal precautions preclude the reduction of glove usage, exposure to NRL and the concomitant risk of sensitization can be expected to grow. Containing the problem requires not only a commitment to the diagnosis and treatment of the

150

TABLE 4 Irritants and Type I and Type IV Allergens in Hypoallergenic Gloves

Brand name	Thiurams	Carbamates	Benzothiazoles	Thiourea	Latex protein	Powder	Company	Phone no.
Surgical								
Puritee Pur	No	No	No	No	Yes	Yes	Puritee Medical Co.	040-682881-0
Eudermic®	No	Yes	No	No	Yes	Yes	Becton Dickinson	(800) 333-4813
Brown Milled	No	No	No	No	Yes	Yes	Professional Medical Products, Inc.	(800) 845-4571
Ultraderm™	No	Yes	No	No	Yes	Yes	Baxter Healthcare	(800) 766-1077
Sempermed	No	Yes	Yes	No	Yes	Yes	Semperit, Inc.	(800) 631-2566
Safeskin	No	Yes	No	No	Yes	Yes	Safeskin Corp.	(800) 456-8379
Microtouch™	No	Yes	Yes	No	Yes	Yes	J & J	(800) 433-5009
Dermaguard+®	No	Yes	No	No	Yes	Yes	Smith & Nephew/Perry	(800) 321-9752
Manex	No	Yes	No	No	Yes	Yes	Beiersdorf, Inc.	011-4-940-5690
Powder Free								
Encore	No	Yes	No	No	Yes	No	Smith & Nephew/Perry	(800) 321-9752
Sempermed Ultra	No	Yes	No	No	Yes	No	Semperit, Inc.	(800) 631-2566
Biogel®	No	Yes	No	No	Yes	No	Regent Hospital Products	(800) 843-8494
Perry Natural™	No	Yes	No	No	Yes	No	Smith & Nephew/Perry	(800) 321-9752
Pristine™	Yes	No	No	No	Yes	No	World Medical Supply	(800) 545-5475
Synthetic								
Elastyren®	No	Yes	No	No	No	Yes	Hermal Pharmaceutical Laboratories	(800) 437-6251
Neolon®	No	Yes	No	No	No	Yes	Becton Dickinson	(800) 333-4813
Dermaprene	No	No	No	Yes	No	Yes	Ansell	(800) 327-8659
Tactyl-1™	No	No	No	No	No	Yes	SmartPractice	(800) 822-8956
Neoderm	No	No	No	No	No	Yes	Beiersdorf, Inc.	011-4-940-5690
Allergard	No	No	No	No	No	Yes	J & J	(800) 433-5009
Examination								
Ansell EP	No	Yes	Yes	No	Yes	Yes	Ansell	(800) 327-8659
Ansell Conform (sterile & nonsterile)	No	Yes	Yes	No	Yes	Yes	Ansell	(800) 327-8659

151

Product						Manufacturer	Phone
Qualitouch	No	Yes	No	No	Yes	SmartPractice	(800) 822-8956
Sempermed EG	No	Yes	No	No	Yes	Semperit, Inc.	(800) 631-2566
Microtouch EG™ (sterile & nonsterile)	No	Yes	Yes	No	Yes	J & J	(800) 433-5009
Dash™	NA	NA	NA	NA	Yes	Dash Medical	(414) 421-2229
Cranberry	NA	NA	NA	NA	Yes	Zeukerman Dental Products	(800) 323-3665
Sensi-Derm®	NA	NA	NA	NA	Yes	Ansell	(800) 327-8659
Powder Free							
Biogel D®	No	Yes	No	No	Yes	Regent Hospital Products	(800) 843-8494
Neutraderm™	No	Yes	Yes	No	Yes	Aladan	(404) 840-9665
Perry Natural™	No	Yes	No	No	Yes	Smith & Nephew/Perry	(800) 321-9752
Royal Shield™	No	Yes	No	No	Yes	SmartPractice	(800) 822-8956
Synthetic							
Vinylite™	No	No	No	No	Yes	SmartPractice	(800) 822-8956
Sensicare	No	No	No	No	Yes	Becton Dickinson	(800) 333-4813
Skincare	No	No	No	No	Yes	Skin Care	(800) 284-8828
Soft Touch	No	No	No	No	Yes	Med Source	(818) 308-0900
Qualitouch™	No	No	No	No	Yes	SmartPractice	(800) 822-8956
Travenol	No	No	No	No	Yes	Baxter Healthcare	(800) 766-1077
Tru-Touch	No	No	No	No	Yes	Becton Dickinson	(800) 333-4813
Excel	No	No	Yes	No	Yes	Veritex	(800) 521-0661
N-Dex	No	No	No	No	Yes	Best	(800) 241-0323
Tactyl-1™	No	No	No	No	Yes	SmartPractice	(800) 822-8956
Dalhausen	No	No	No	No	No	P.J. Dahlhausen[a]	NA
4H glove®	No	No	No	No	No	Safety 4[b]	NA

Note Data subject to change.

[a] In Cologne, FRG; phone number not available.
[b] In Lyngby, Denmark.

existing sensitized population, but research on how to prevent future sensitization of new individuals.

Research aimed at decreasing the incidence of NRL sensitization is proceeding along two main fronts. The first approach is concentrating on reducing or eliminating the NRL allergen(s) from gloves while maintaining the physical strength and effectiveness of the barrier properties. At the International Latex Conference convened in November 1992 by the FDA Center for Devices and Radiological Health, the Centers for Disease Control, and the National Institute for Allergy and Infectious Diseases, representatives from industry analyzed production methods that could be used to reduce the content of protein allergens in finished devices.[65] Various methods can decrease the level of extractable proteins during preprocessing, the actual manufacture, and postprocessing: increased ammoniation, multiple centrifugation, treatment with proteolytic enzymes, increased leaching with water or chlorine, autoclaving, and the surface application of silicone to products.[5] Unfortunately, each of these manipulations can affect the equally important barrier and physical properties of a glove. Consequently, it is unlikely that NRL allergens can be completely eliminated from products in the near future.

Considerable data are accumulating in the second major area of research — identification of the specific NRL protein allergen(s). Nonetheless, variability in technique and antigen source has hampered interpretation of the results. A spectrum of proteins, ranging from 2 to 200 kDa, has been implicated, but their clinical relevance has not been established.[24-33] More intensive collaboration among researchers in this area, with pooled sera and standardized techniques, would likely hasten allergen identification.[12]

Progress along either of these lines would simplify the task of diagnosis. Clear identification of allergens, especially if combined with product knowledge (enhanced by labeling), would narrow the treatment choices by eliminating many brands from consideration by the patient with defined glove sensitivities. Even more desirable, identification of the specific allergen could be combined selectively with production methods targeted to reduce that protein(s). A new generation of gloves with lower concentrations of allergens and irritants could be developed in response to such breakthroughs. The continued cooperative effort between industry and the medical research community offers the best hope for insuring that protective gloves protect, rather than harm, workers.

REFERENCES

1. Downing, J. G., Dermatitis from rubber gloves, *N. Engl. J. Med.,* 208, 196, 1933.
2. Nutter, A. F., Contact urticaria to rubber, *Br. J. Dermatol.,* 101, 597, 1979.
3. Berky, Z. T., Lucianc, W. J., and James, W. D., Latex glove allergy: a survey of the US Army dental corps, *J. Am. Med. Assoc.,* 268, 2695, 1992.

4. Bason, M., Lammintausta, K., and Maibach, H. I., Irritant dermatitis (irritation), in *Dermatotoxicology,* 4th ed., Marzulli, F. N. and Maibach, H. I., Eds., Hemisphere, New York, 1991, 223.
5. Truscott, W., Manufacturing Methods Sought to Eliminate or Reduce Sensitivity to Natural Rubber Products, in *Prog. Proc. Int. Latex Conf. Sensitivity to Latex in Medical Devices,* U.S. Food and Drug Administration, Washington, D.C., 1992, 66.
6. Heese, A., Hintzenstern, J. V., Peters, K. P., et al., Allergic and irritant reactions to rubber gloves in medical health services. Spectrum, diagnostic approach, and therapy, *J. Am. Acad. Dermatol.,* 25, 831, 1991.
7. Hamann, C., Latex/rubber allergies in the health care environment. Part II. High fiber, low fat, hypoallergenic surgical gloves? The problem with "hypoallergenic" gloves, *Contemp. Dialysis. Nephrol.,* 13, 2, 1992.
8. Shmunes, E. and Darby, T., Contact dermatitis due to endotoxin in irradiated latex gloves, *Contact Derm.,* 10, 240, 1984.
9. Fisher, A. A., Burns on the hands due to ethylene oxide used to sterilize gloves, *Cutis,* 42, 267, 1988.
10. Coombs, R. R. A. and Gell, P. G. H., The classification of allergic reactions underlying disease, in *Clinical Aspects of Immunology,* Gell, P. G. H. and Coombs, R. R. A., Eds., F. A. Davis, Philadelphia, 1963, 317.
11. Taylor, J. S., Contact dermatitis from rubber gloves, in *Contact Dermatitis,* 3rd ed., Fisher, A. A., Ed., Lea & Febiger, 1986, 631.
12. Hamann, C. P., Natural rubber latex protein sensitivity in review, *Am. J. Cont. Derm.,* 4, 4, 1993.
13. Layer, R. W. and Lattimer, R. P., Protection of rubber against ozone, *Rubber Chem. Technol.,* 63, 426, 1990.
14. Brandão, F. M., Rubber, in *Occupational Skin Disease,* 2nd ed., Adams, R. M., Ed., W. B. Saunders, Philadelphia, 1990, 462.
15. Cronin, E., Rubber, in *Contact Dermatitis,* Cronin, E., Ed., Churchill Livingstone, Edinburgh, 1980, 714.
16. Hintzenstern, J., Heese, A., Koch, H. U., et al., Frequency, spectrum and occupational relevance of type IV allergies to rubber chemicals. A retrospective study from the Department of Dermatology, University of Erlangen-Nuremberg 1/1985–3/1990, *Contact Derm.,* 24, 244, 1991.
17. van Ketel, W. G. and van den Berg, W. H., The problem of the sensitization to dithiocarbamates in thiuram-allergic patients, *Dermatologica,* 169, 70,1984.
18. Frosch, P. J., Wahl, R., Bahmer, F. A., et al., Contact urticaria to rubber gloves is IgE-mediated, *Contact Derm.,* 14, 241, 1986.
19. Turjanmaa, K., Reunala, T., Alenius, H., et al., Allergens in latex surgical gloves and glove powder, *Lancet,* 336, 1588, 1990.
20. Esau, K., *Plant Anatomy,* 2nd ed., John Wiley & Sons, New York, 1965, 318.
21. Pendle, T. D., The Production, Composition and Chemistry of Natural Latex Concentrates, in *Prog. Proc. Int. Latex Conf. Sensitivity to Latex in Medical Devices,* U.S. Food and Drug Administration, Washington, D.C., 1992, 13.
22. Dennis, M. S., Henzel, W. J., Bell, J., et al., Amino acid sequence of rubber elongation factor protein associated with rubber particles of *Hevea* latex, *J. Biol. Chem.,* 264, 18618, 1989.

23. Archer, B. L., The proteins of *Hevea brasiliensis* latex, *Biochem. J.,* 75, 236, 1960.
24. Mäkinen-Kiljunen, S., Turjanmaa, K., Palosuo, T., et al., Characterization of latex antigens and allergens in surgical gloves and natural rubber by immunoelectrophoretic methods, *J. Allergy Clin. Immunol.,* 90, 230, 1992.
25. Alenius, H., Turjanmaa, K., Palosuo, T., et al., Surgical latex glove allergy. Characterization of rubber protein allergens by immunoblotting, *Int. Arch. Allergy Appl. Immunol.,* 96, 376, 1991.
26. Morales, C., Basomba, A., Carreira, J., et al., Anaphylaxis produced by rubber glove contact. Case reports and immunological identification of the antigens involved, *Clin. Exp. Allergy,* 19, 425, 1989.
27. Turjanmaa, K., Laurila, K., Makinen-Kiljunen, S., et al., Rubber contact urticaria. Allergenic properties of 19 brands of gloves, *Contact Derm.,* 19, 362, 1988.
28. Turjanmaa, K. and Reunala, T., Condoms as a source of latex allergen and cause of contact urticaria, *Contact Derm.,* 20, 360, 1989.
29. Fuchs, T. and Wahl, R., Immediate reactions to rubber products, *Allergy Proc.,* 13, 61, 1992.
30. Jaeger, D., Kleinhans, D., Czuppon, A. B., et al., Latex-specific proteins causing immediate-type cutaneous, nasal, bronchial, and systemic reactions, *J. Allergy Clin. Immunol.,* 89, 759, 1992.
31. Slater, J. E. and Chhabra, S. K., Latex antigens, *J. Allergy Clin. Immunol.,* 89, 673, 1992.
32. Zacharisen, M. C., Kurup, V. P., and Resnick, A., Characterization of latex antigens reacting with Ig-E antibodies in the sera of patients with allergy to latex, *J. Allergy Clin. Immunol.,* 89, 225, 1992.
33. Mathew, S. N., Melton, A., Wagner, W., et al., Latex hypersensitivity. Prevalence among children with spina bifida and immunoblotting identification of latex proteins, *J. Allergy Clin. Immunol.,* 89, 225, 1992.
34. FDA, Allergic reactions to latex-containing medical devices, *FDA Medical Bulletin,* U.S. Food and Drug Administration, Washington, D.C., July, 1991.
35. Centers for Disease Control, Anaphylactic reactions during general anesthesia among pediatric patients — United States, January 1990–January 1991, *MMWR,* 40, 437, 1991.
36. Ownby, D. R., Tomlanovich, M., Sammons, N., et al., Anaphylaxis during barium enema examinations associated with latex allergy, *J. Allergy Clin. Immunol.,* 156, 903, 1991.
37. Turjanmaa, K. and Reunala, T., Contact urticaria from rubber gloves, *Dermatol. Clin.,* 6, 47, 1988.
38. Kwittken, P. L. and Sweinberg, S. H., Childhood latex allergy. An overview, *Am. J. Asthma Allergy Ped.,* 6, 27, 1992.
39. Medical Alert Bulletin, FDA, U.S. Food and Drug Administration, Washington, D.C., March 29, 1991.
40. Turjanmaa, K., Incidence of immediate allergy to latex gloves in hospital personnel, *Contact Derm.,* 17, 270, 1987.
41. Tarlo, S. M., Wong, L., Roos, J., et al., Occupational asthma caused by latex in a surgical glove manufacturing plant, *J. Allergy Clin. Immunol.,* 85, 626, 1990.
42. Slater, J. E., Mostello, L. A., Shaer, C., et al., Type I hypersensitivity to rubber, *Ann. Allergy,* 65, 411, 1990.

43. Meeropol, E., Kelleher, R., Bell, S., et al., Allergic reactions to rubber in patients with myelodysplasia, *N. Engl. J. Med.,* 323, 1072, 1990.
44. Young, M. A., Meyers, M., McCulloch, L. D., et al., Latex allergy. A guideline for perioperative nurses, *AORN J.,* 56, 488, 1992.
45. Bubak, M. E., Reed, C. E., Fransway, A. F., et al., Allergic reactions to latex among health-care workers, *Mayo Clin. Proc.,* 67, 1075, 1992.
46. Axelsson, J. G. K., Johansson, S. G. O., and Wrangsjö, K., Ig-E mediated anaphylactoid reactions to rubber, *Allergy,* 42, 46, 1987.
47. Estlander, T., Jolanki, R., and Kanerva, L., Dermatitis and urticaria from rubber and plastic gloves, *Contact Derm.,* 14, 20, 1986.
48. Förström, L., Contact urticaria from latex surgical gloves, *Contact Derm.,* 6, 33, 1980.
49. De Corres, L., Moneo, I., Muñoz, D., et al., Sensitization from chestnuts and bananas in patients with urticaria and anaphylaxis from contact with latex, *Ann. Allergy,* 70, 35, 1993.
50. M'Raihi, L., Charpin, D., Pons, A., et al., Cross-reactivity between latex and banana, *J. Allergy Clin. Immunol.,* 87, 129, 1991.
51. Young, M. C., Osleeb, C., and Slater, J., Latex and banana anaphylaxis, *J. Allergy Clin. Immunol.,* 89, 226, 1992.
52. Lagier, F., Vervloet, D., Lhermet, I., et al., Prevalence of latex allergy in operating room nurses, *J. Allergy Clin. Immunol.,* 90, 319, 1992.
53. Ceuppens, J. L., Van Durme, P., and Dooms-Goossens, A., Latex allergy in patient with allergy to fruit, *Lancet,* 339, 493, 1992.
54. Fisher, A. A., Association of latex and food allergy, *Cutis,* 52, 70, 1993.
55. Fisher, A. A., Allergic contact reactions in health personnel, *J. Allergy Clin. Immunol.,* 90, 729, 1992.
56. Maso, M. J. and Goldberg, D. J., Contact dermatoses from disposable glove use. A review, *J. Am. Acad. Dermatol.,* 23, 733, 1990.
57. Dooms-Goossens, A., Contact urticaria from rubber gloves, *J. Am. Acad. Dermatol.,* 18, 1360, 1988.
58. Bradley, J., Sussman, G. L., and Arellane, R., Incidence of immediate allergy to latex gloves in 100 hospital physicians at the University of Toronto, *Can. J. Anaestheol.,* 38, A100, 1991.
59. Sussman, G. L., Safety of latex prick skin testing in allergic patients (reply), *J. Am. Med. Assoc.,* 267, 2603, 1992.
60. von Blomberg, B. M. E., Bruynzeel, D. P., and Scheper, R. J., Advances in mechanisms of allergic contact dermatitis: *in vitro* and *in vivo* research, in *Dermatotoxicology,* 4th ed., Marzulli, F. N. and Maibach, H. I., Eds., Hemisphere, New York, 1991, 225.
61. Hjorth, N., Diagnostic patch testing, in *Dermatotoxicology,* 4th ed., Marzulli, F. N. and Maibach, H. I., Eds., Hemisphere, New York, 1991, 441.
62. Mitchell, J. C., Clendenning, W. E., Cronin, E., et al., Patch testing with mercaptobenzothiazole and mercapto-mix, *Contact Derm.,* 2, 123, 1976.
63. Turjanmaa, K. and Reunala, T., Contact urticaria to surgical and household rubber gloves, in *Exogenous Dermatoses: Environmental Dermatitis,* Menne, T. and Maibach, H. I., Eds., CRC Press, Boca Raton, FL, 1991, 317.
64. Spaner, D., Dolovich, J., Tarlo, S., et al., Hypersensitivity to natural latex, *J. Allergy Clin. Immunol.,* 83, 1135, 1989.

65. Taylor, J. S., Melton, A., and Hamann, C., Selected highlights of the International Latex Conference. Sensitivity to latex in medical devices, Baltimore, MD, November 5–7, 1992, *Am. J. Contact Derm.,* 4, 1, 1993.
66. Stites, D. P. and Terr, A. I., *Basic and Clinical Immunology,* 6th ed., Appleton & Lange, Norwalk, CT, 1987, 373.
67. Von Krogh, G. and Maibach, H. I., The contact urticaria syndrome, *J. Am. Acad. Dermatol.,* 8, 328, 1981.
68. Turjanmaa, K., Reunala, T., and Räsänen, L., Comparison of diagnostic methods in latex surgical glove contact urticaria, *Contact Derm.,* 19, 241, 1988.
69. Pecquet, C., Leynadier, F., and Dry, J., Contact urticaria and anaphylaxis to natural latex, *J. Am. Acad. Dermatol.,* 22, 631, 1990.
70. Yassin, M. S., Sanyurah, S., Lierl, M. B., et al., Evaluation of latex allergy in patients with meningomyelocele, *Ann. Allergy,* 69, 207, 1992.
71. Losada, E., Lazaro, M., Marcos, C., et al., Immediate allergy to natural latex: clinical and immunological studies, *Allergy Proc.,* 13, 115, 1992.
72. Beezhold, D. H., LEAP: latex ELISA for antigenic proteins. Preliminary report, *Guthrie J.,* 61, 77, 1992.
73. Turjanmaa, K., Räsänen, L., Lehto, M., et al., Basophil histamine release and lymphocyte proliferation tests in latex contact urticaria, *Allergy,* 44, 181, 1989.
74. Carrillo, T., Cuevas, M., Muñoz, T., et al., Contact urticaria and rhinitis from latex surgical gloves, *Contact Derm.,* 15, 69, 1986.
75. Marzulli, F. N. and Maibach, H. I., Contact allergy: predictive testing in humans, *Dermatotoxicology,* 4th ed., Marzulli, F. N. and Maibach, H. I., Eds., Hemisphere, New York, 1991, 415.
76. Beezhold, D. and Beck, W. C., Surgical glove powders bind latex antigens, *Arch. Surg.,* 127, 1354, 1992.
77. Korniewicz, D. M., Laughon, B. E., Butz, A., et al., Integrity of vinyl and latex procedure gloves, *Nurs. Res.,* 38, 144, 1989.
78. Korniewicz, D. M., Laughon, B. E., Cyr, W. H., et al., Leakage of virus through used vinyl and latex examination gloves, *J. Clin. Microbiol.,* 28, 787, 1990.
79. Kotilainen, H. R., Avato, J. L., and Gantz, N. M., Latex and vinyl nonsterile examination gloves. Status report on laboratory evaluation of defects by physical and biological methods, *Appl. Environ. Microbiol.,* 56, 1627, 1990.

11

Clinical Testing for Rubber Glove Allergy

Michael H. Beck and Christopher R. Lovell

TABLE OF CONTENTS

I. Introduction 157
II. Contact Urticaria 158
III. Allergic Contact Dermatitis 161
    A. Standard Patch Tests 161
    B. Analysis of Results 162
    C. Comments 163
        1. False Negative Reactions 165
        2. False Positive Reactions 165
IV. Occupation and Rubber Glove Allergy 167
    A. Manchester Results 167
    B. The Occ-Derm Project 167
V. Conclusion 168
Acknowledgments 168
References 169

I. INTRODUCTION

In the U.K., clinical tests for contact allergy are generally performed by individual dermatologists, or patients may be referred to a specialized contact dermatitis investigation unit. The work presented here is from two centers. In Bath, one of us (C. R. Lovell) has performed research on aspects of immediate type I hypersensitivity from rubber gloves, presenting clinically as contact urticaria. In Manchester, the results from an investigation clinic looking at type IV delayed hypersensitivity presenting clinically as contact dermatitis are presented with particular reference to rubber gloves.

The majority of rubber imports to the U.K. come from Malaysia, which provides two thirds of the world's concentrated rubber latex production. Once the rubber has been tapped, the latex is ammoniated to prevent microbial degradation and concentrated before storage. In the production of rubber gloves, latex is vulcanized and a wide range of chemicals may be added.

Contact urticaria from rubber gloves is caused by latex protein, but the vast majority of allergic contact dermatitis relates to the chemicals added, in particular, accelerators and antioxidants.

II. CONTACT URTICARIA

It is our impression that contact urticaria is not as common in the U.K. compared with some other regions, notably Scandinavia. This may, however, be a reflection of referral patterns to dermatologists in our country, rather than a genuine observation. Manufacturing processes might play a part, as there is variability in the amount of extractable latex protein in rubber gloves.[1]

Latex immediate hypersensitivity presents a potentially hazardous problem to affected health care professionals and patients during surgical procedures. In consequence, the rubber industry is studying ways to reduce the amount of extractable latex protein in the gloves.

Extractable proteins are dispersed irregularly on the surface of the gloves.[1] Further washing of gloves before wearing may prevent urticarial reactions in sensitized individuals.[2] Ideally, this process should be carried out during the industrial manufacture of the glove; allergenicity being reduced by washing the gloves after mold forming and then subjecting them to steam sterilization.[3]

In conjunction with Dr. Brian Audley and colleagues at the Tun Abdul Razak Laboratory, we have studied the effect of different extraction techniques on the allergenicity of rubber latex. Four patients were studied between 1990 and 1992, three females (ages 28 to 36) and one male (aged 23). The case history of patient 1 is illustrative. She presented as a medical emergency in anaphylactic shock 1 h after a vaginal examination by a gynecologist using rubber gloves. She made a rapid recovery following subcutaneous adrenaline. Subsequently, she developed a further anaphylactic attack after blowing up rubber balloons at a children's party and also developed episodes of eyelid swelling after entering rooms containing rubber gloves or balloons. The other three patients, all nurses, described swelling and itching of the hands, associated with generalized weals, after wearing latex rubber gloves. One (patient 2) observed eyelid swelling when she walked into the endoscopy suite, where rubber gloves were used.

All patients were prick test positive to ammoniated natural latex. Note that *in vivo* testing needs to be conducted with caution. Favored techniques include a "use test" with a finger cut from a rubber glove, or open application of a piece of glove to the dorsum of the hand, perhaps after wetting the skin. Even wearing a glove for 10 min has induced bronchospasm and widespread weals in a sensitized hospital nurse.[4] Only after these tests have been performed

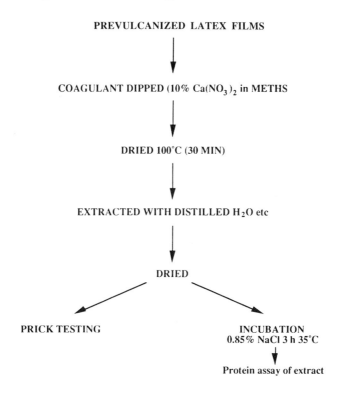

FIGURE 1 Dipping and drying process undergone for latex films.

should one consider prick or closed patch testing and resuscitation facilities should be present.

In order to quantify the association between allergenicity and extractable protein content, rubber films were subjected to different extraction techniques at the Tun Abdul Razak Laboratory. All the latex films were coagulant dipped and dried (Figure 1) before the extraction procedure. Four samples of film were then washed in distilled water with agitation for 1 to 60 min (Table 1). Wet gel leaching was performed on three further samples of latex film using static distilled water for 2 min at 60 or 80°C. Two of these three samples were further treated. One was steam-treated at 120°C for 1 h and another sample was chlorinated according to the procedure outlined in Table 2. A portion of each sample was returned in order to measure the degree of extractable protein using the Bradford method.[5] The results are presented in Table 1. Further details of the extraction procedures are given in the paper by Dalrymple and Audley.[1] Each patient was then prick tested using a 0.5 × 0.5 cm portion of the final film on a "blind" basis. The prick test results are given in Table 3.

It can be seen that prick test size correlates roughly with the amount of extractable protein remaining in the film. Accepting that wheal sizes of less

TABLE 1 Latex Extraction Procedures and Residual Extractable Protein Levels

	Extraction procedure	Time (min)	Temperature (°C)	Further treatment	Extractable protein (µg/g rubber)
Sample 1	Dry film	1	25		230
2	washed in	5	25		140
3	agitated	15	25		130
4	distilled water	60	25		70
5	Wet gel leached	2	80		315
6	in static	2	60	Steam 120°C (1 h)	125
7	distilled water	2	60	Chlorination	<35

TABLE 2 Chlorination Procedure

Immerse dried film in 0.3% aqueous solution of chlorine — 3 min
Rinse in distilled water — 1 min × 2
Neutralize residual acid 1% aqueous ammonia solution — 1 min
Rinse in distilled water — 1 min

TABLE 3 Results of Prick Tests With Post-Extraction Latex Films

Patients	Prick test reactions (wheal mm)				Extractable protein (mg/g rubber)
	1	2	3	4	
Sample 5	16	13	13	15	315
Sample 1	18	14	13	13	230
Sample 2	5	9	10	8	140
Sample 3	7	9	11	4	130
Sample 6	—	10	8	5	125
Sample 4	—	6	3	3	70
Sample 7	—	—	3	—	<35

than 4 mm are unlikely to be of significance, it can be seen that adequate washing in distilled water for 1 h gave an acceptable preparation for 3 of the 4 subjects. However, chlorination proved the most effective technique, both in terms of extractable protein yield and lack of allergenicity on prick testing. This may be due to associated treatments such as exposure to water, salt, or ammonia, rather than chlorination itself.[1] Unfortunately, this is a complex process unlikely to be popular with the rubber industry.

In conclusion, our study supports the generally held view that the allergens are water extractable proteins present in both the prevulcanized latex and on the finished product. As the antigenic protein has not been identified precisely, it is clearly desirable to lower the total content of extractable protein. Prior washing of surgical gloves and other rubber equipment is not always practicable before use, and numerous extraction techniques are currently under investigation in the rubber industry.[1,3]

It is not clear why type I hypersensitivity is a relatively recent phenomenon. There has been an increased demand for rubber gloves and condoms as a protection against HIV infection. It is possible that the increased requirement has resulted in some manufacturers reducing the extraction procedure, resulting in finished products with a higher concentration of extractable proteins. Alternatively, different clones of *Hevea brasiliensis* may yield latex of different antigenicity, or stimulation of tree growth may affect the chemical composition of the latex. Additionally, once a condition is described then further reports inevitably will follow. The relatively paucity of reported cases in the U.K. as compared with Scandinavia remains conjectural.

III. ALLERGIC CONTACT DERMATITIS

The Skin Hospital Contact Dermatitis Investigation Unit at Manchester investigates persons referred by dermatologists from a catchment population of 2,750,000. The results of these investigations, with particular reference to rubber gloves, will be presented. The number we are able to see in this clinic will not reflect the true incidence of rubber glove dermatitis in this population, because all patients with hand dermatitis are not referred by their own doctor to a dermatologist. Furthermore, dermatologists may not necessarily refer all their patients to our unit and some do their own patch testing. Additionally, some affected individuals recognize the cause of the problem themselves and may not even seek medical advice.

A. Standard Patch Tests

All patients referred to the unit are tested with a standard series of common contact allergens in our environment, usually amounting to 30 to 40 allergens in all. In this battery of tests are included four markers of allergens potentially found in rubber gloves. They are currently provided for us by Hermal Chemie. The allergens are as follows:

Mercaptobenzothiazole (MBT)	1% in petrolatum
Mercapto mix	2% in petrolatum
Carba mix	3% in petrolatum
Thiuram mix	1% in petrolatum

The ingredients of the mixes are shown in Table 4.

Patients are also tested to additional allergens of relevance and other materials brought by them, at non-irritant concentrations. If rubber gloves are worn at any time in the domestic or working environment, then we ask the subjects to bring them and a 2-cm square is removed for testing.

The allergens are normally applied to the upper back using aluminum chambers and left for 48 h. They are then removed and any reaction noted and

TABLE 4 Standard Series Rubber Allergens: Components of "Mixes"

Mercapto mix	
Mercaptobenzothiazole	MBT
N-cyclohexyl-benzothiazol sulphenamide	CBS
Morpholinyl-mercaptobenzothiazole	MOR
Dibenzothiazyl disulfide	MBTS
Each at 0.5%	
Carba mix	
Bis (diethyldithiocarbamate) zinc	ZDC
Bis (dibutyldithiocarbamate) zinc	ZBC
Diphenylguanidine	DPG
Each at 1%	
Thiuram mix	
Tetramethyl thiuram disulfide	TMTD
Tetraethyl thiuram disulfide	TETD
Tetramethyl thiuram monosulfide	TMTM
Dipentamethylene thiuram disulfide	PTD
Each at 0.25%	

read, according to standard ICDRG recommendations.[6] A further reading is made 48 h later. Thereafter, an interpretation is made with regard to relevance. Readings have all been performed by dermatologists working in the clinic.

B. Analysis of Results

A total of 6720 patients referred to our clinic have been analyzed. Of these, 2112 (31.9%) had their primary site of involvement recorded as the hands.

A total of 152 individuals were felt to have an allergic contact dermatitis to one or more of the rubber allergens, due to their presence in rubber gloves. This amounts to 2.3% of all patients seen and 7.2% of those with hand dermatitis. Note that these figures do not account for all our rubber sensitive individuals, as others were sensitized from different sources and in other sites, e.g., footwear, elastic in clothing, and other rubber contact by the hands. Of the 152 affected, 92 were female (60.5%). There was an age range of 14 to 88. The age distribution is shown in Figure 2. A breakdown of the reactions for individual allergen groups can be seen in Tables 5 and 6.

If we analyze those causes of allergic contact dermatitis of the hands interpreted as being of some relevance by the investigator, then rubber gloves, as identified by the history, examination, and these four test allergens, are the fourth most common cause of such sensitization (Table 7). It is, however, perhaps an artificial comparison as sources for the other allergens may be many. Furthermore, the relevance of nickel, cobalt, and fragrance sensitivity to hand eczema may be impossible to quantify, particularly as our scoring system includes possible (or dubious) relevance. This may mean that these materials are overrepresented in the table. Generally, one can be rather more certain about the relevance of the rubber chemicals.

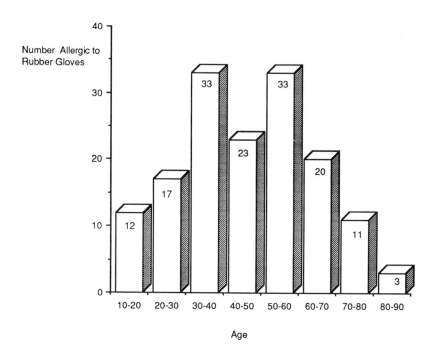

FIGURE 2 Age range of rubber glove allergic subjects.

It is possible to analyze these figures in more detail, particularly with regard to concomitant sensitization and cross-sensitization, and these figures are shown in Tables 8, 9, and 10. Of the 152 patients allergic to rubber chemicals found in gloves, 103 were tested with their own gloves and 56 (54.4%) had reactions at patch testing interpreted as allergic. Additionally, four had strong ++ reactions and negative tests to our rubber chemical screen. Attempts to test our patients further with all the ingredients of the materials from the gloves failed, as two persons were unable to attend for the extra tests and two sets of gloves were made abroad and we could not obtain all the ingredients. False positive reactions could not be ruled out completely.

A further 420 patients negative to rubber chemicals have been tested with rubber gloves and 62 (13.2%) had + or +/− reactions, particularly at the first reading interpreted as false positive irritant reactions due to pressure, especially around the edge of the test sample.

C. Comments

From our results, it would appear that screening for allergic contact dermatitis from rubber gloves is satisfactorily covered, in general, by a combination of

TABLE 5 Standard Series Rubber Allergens Results in 152 Glove Allergic Subjects

	Total	% Hand dermatitis (2112 patients)	% Total (6720 patients)
MBT	35	1.66	0.52
Mercapto mix	29	1.37	0.43
Carba mix	77	3.65	1.15
Thiuram mix	109	5.16	1.62

TABLE 6 Standard Series Rubber Allergens Results in 152 Glove Allergic Subjects According to Sex

	Mercaptans	Carba mix	Thiuram mix
Males	13 (33.3%)	30 (36%)	42 (39%)
Females	26 (66.6%)	47 (64%)	67 (61%)

TABLE 7 Causes of Relevant Contact Sensitization in 2112 Patients With Hand Dermatitis

Nickel	19.7%
Fragrances	9.4%
Cobalt	7.7%
Rubber gloves	7.2%
Colophony	6.1%
Formaldehyde	3.1%
Neomycin	2.5%
Lanolins	2.4%
Quaternium 15	2.1%

TABLE 8 Patients With Mercaptan Allergy from Rubber Gloves

	MBT alone	Mercapto mix alone	Both MBT and mercapto mix	Total
Males	2	0	11	13
Females	8	4	14	26
Total	10	4	25	39

TABLE 9 Patients With Thiuram and Carbamate Allergy from Rubber Gloves

	Carba mix alone	Thiuram mix alone	Both thiuram and carba mix	Total
Males	10	22	20	52
Females	8	28	39	75
Total	18	50	59	127

TABLE 10 Concomitant Allergy to Mercaptans, Carba, and Thiuram Mixes

	Mercaptans alone	Mercaptans and thiuram and/or carba mixes	Total
Males	8	5	13
Females	17	9	26
Total	25	14	39

mercaptobenzothiazole, mercapto mix, carba mix, and thiuram mix in the standard series of patch test allergens. However, the rubber gloves worn by the individuals should also be tested. There are problems, however, with testing with the gloves, which are as follows.

1. False Negative Reactions

In our series of those tested, 45.6% had false negative reactions to the glove. The reasons for this may be varied. Firstly, not all the gloves worn may have been brought for testing. Secondly, people might change the brand of glove from time to time when they buy a new pair and the composition may vary from glove to glove. Much more likely, however, is that insufficient allergen is released by the rubber from the glove to induce a positive reaction, by virtue of the small amount of allergen in the glove in contact with normal skin. It must be remembered that allergic contact dermatitis to rubber commonly occurs in association with an irritant contact dermatitis which, if preexisting, will have damaged the skin, thus assisting allergen penetration and sensitization.

2. False Positive Reactions

Although rubber from a glove is flexible, it is still sufficiently rigid to have a direct pressure effect on the skin, particularly if an edge is pressed down by the test chamber. In consequence, particularly at the time of removal of the test units, there may be seen an erythema sufficiently strong to record as a + reaction, particularly if the individual is dermographic, and perhaps even a ++ reaction may be recorded at the first reading. Fortunately, in most cases this reaction has largely disappeared by the second reading. However, some persistent erythema may still be noted as a result of the direct damage of pressure and friction on the skin caused by the rubber sample. This may be particularly prominent around the edge, where there will be an erythematous linear configuration.

The increased number (60.5%) of females, compared with males might be expected as exposure to rubber gloves is more likely. However, more females were tested (60.2%), so these figures would suggest a roughly equal sex distribution in those who were referred to our clinic.

Of the screening allergens, thiuram mix yielded the highest return in rubber glove allergic subjects, amounting to 71.7% of such persons. Just over half of these were allergic to carba mix as well. False positive irritant reactions to thiuram mix were rare. Weak +/− reactions were not interpreted, but 27 reactions out of the 6760 persons tested occurred which were recorded as + at 48 or 96 h, which were felt to be false positives, this being reinforced by negative reactions when the individual components were tested later.

Carba mix identified just over half the glove-sensitive individuals. We have, however, found problems patch testing with this particular mix, as false positive reactions are particularly common, and in many cases the morphology and behavior of the reaction is in our view indistinguishable from an allergy one.

The situation is further complicated by the fact that patch testing with one of the breakdown materials (DPG) can similarly give false positive reactions. A + reaction to carba mix and DPG with a negative reaction to thiuram mix can cause much deliberation by the patch test investigator! This conundrum is further complicated by the diametrically opposed views in the literature on the subject of an isolated carba mix reaction.[7,8] The consensus among the EECDRG was to concur with the view of Logan and White[7] and advise that carba mix be dropped from the standard series.[9]

Our results do not satisfactorily resolve the difficulty. We have identified 82 individuals with + or ++ false positive reactions out of the 6760 tested. Additionally, 141 persons were recorded with 2- or 4-day readings as +/− and no interpretation was made. We have assumed that these are also false positive irritant reactions.

The view which we take is that where we see a + or ++ reaction to carba mix and one or more of its ingredients in a thiuram mix negative individual who wears rubber gloves, and who has hand eczema, is that we advise the patient that we believe he/she is allergic to rubber and to take appropriate contact avoidance steps. At the same time we accept that we may have interpreted a false positive reaction as truly allergic. At this time we certainly feel happier in maintaining carba mix in the standard series, but wonder whether the concentration of DPG might be dropped in the mix.

MBT and mercapto mix allergy accounts for just over a quarter of our cases with glove allergy. It is now recognized that both MBT and mercapto mix need to be incorporated in the standard series and our results would reinforce this view, as ten cases of MBT allergy would have been missed had we relied on mercapto mix alone, even though it contains MBT, but at a concentration of 0.5% as against 1% when tested individually. False positive reactions were not a problem. We are not aware of any attempt to look at the feasibility and reliability of increasing the MBT concentration in mercapto mix to 1%.

IV. OCCUPATION AND RUBBER GLOVE ALLERGY

We have analyzed information not only from the Manchester cohort of rubber glove allergic patients, but also from the Occ-Derm Project. The latter is a pilot study supervised by the British Contact Dermatitis Research group and funded by the Health and Safety Executive where 16 centers in the U.K. have recorded newly identified occupational dermatoses, intending to look at the feasibility of setting up a national reporting system for such problems.

A. Manchester Results

Of the 152 rubber glove-sensitive individuals, 71 (46.7%) were felt to have relevant occupational exposures (housewives were not included as an occupational group). Only 4 groupings had more than 2 individuals in them, namely: catering with 16, cleaners with 12, medical occupations with 11, and chemical process workers with 5. Rubber gloves are particularly liable to be used in these occupations as protection, so it is perhaps no surprise to see these occupations well represented.

B. The Occ-Derm Project

Over a 2.5-year period 2861 cases of occupational dermatitis were identified by our group. A total of 306 cases (10.7%) were documented as reacting to rubber, rubber gloves, or rubber chemicals, and although rubber gloves were only specifically mentioned in 87 (3%) of these cases, we believe that the vast majority of the total were sensitized to the rubber or rubber chemicals from their gloves, furthermore, 81 (2.8%) were documented as being thiuram sensitive.

Occupations with greater than five persons identified are shown in Table 11. In addition, 12 cases of contact urticaria to latex were identified, each in a separate occupational group. It is likely that medical and allied professions are likely to be overrepresented, compared with other occupations, as they will have greater opportunity for access to dermatologists and patch testing.

The figures, of course, are absolute and do not necessarily represent a relative risk. The increasing need for dental workers, as a result of the increasing prevalence of HIV infection, and the length of time they need to wear the gloves throughout the day, clearly is putting them at risk of sensitization.

A figure of over 10% of all the reported dermatitis cases being due to rubber chemicals, largely through the use of rubber gloves, is of significance bearing in mind the considerable morbidity and financial consequences of occupational dermatitis.

TABLE 11 The Occ-Derm Project — Analysis of Occupational Dermatitis

Occupation	Total with rubber allergens	Total with dermatitis	% Rubber allergy
Nurses	32	120	26.7
Cleaners/domestics	27	133	20.3
Counter hands	19	154	12.3
Catering assistants/kitchen porters	7	39	17.9
Laborers in manufacturing and processing industries	13	161	8.1
Metal working, production, and maintenance fitters	10	114	8.8
Chefs/cooks	8	114	7.0
Dental practitioners	8	14	57.1
Other laborers and related	8	25	32.0
Hairdresser	7	249	2.8
Machine tool operators	7	94	7.4
Dental nurses	6	18	33.3
Chemical, gas and petroleum plant operatives	6	57	10.5

V. CONCLUSION

The relative frequency and consequences of rubber glove allergy in our studies point to the need to continue to assess and reexamine closely the treatment and manufacturing processes performed on rubber when made into gloves in order to see if there are ways of reducing or eliminating the problems identified without affecting the end product.

ACKNOWLEDGMENTS

We wish to thank Dr. Brian Audley and colleagues at the Tun Abdul Razak Laboratory, Hertford, for preparing the latex films and measuring extractable protein levels; Drs. Mansel, Reckless, Heap, and Tanser for referring the patients with contact urticaria; Dr. Valerie Hillier of the Department of Medical Computation, University of Manchester Medical School, for assistance with the analysis of the contact dermatitis investigations; Professor Nicola Cherry and Dr. Victoria Owen-Smith of the University of Manchester Centre for Occupational Health, for the analysis of the Occ-Derm Project; The Medical Illustration Department, Bolton General Hospital, for the figures; and Mrs. Sue Parkinson for preparing and typing the manuscript.

REFERENCES

1. Dalrymple, S. J. and Audley, B. G., Allergenic proteins in dipped products: factors influencing extractable protein levels, *Rubber Dev.,* 45, 51, 1992.
2. Morales, C., Basomba, A., Carreira, J., and Sastre, A., Anaphylaxis produced by rubber glove contact. Case reports and immunological identification of the antigens involved, *Clin. Exp. Allergy,* 19, 425, 1989.
3. Leynadier, F., Tran Xuan, T., and Dry, J., Allergenicity suppression in natural latex surgical gloves, *Allergy,* 46, 619, 1991.
4. DeZotti, R., Larese, F., and Fiorito, A., Asthma and contact urticaria from latex gloves in a hospital nurse, *Br. J. Ind. Med.,* 49, 596, 1992.
5. Bradford, M. M., A rapid and sensitive method for the quantitation of microgram quantities of protein utilising the principle of protein-dye binding, *Anal. Biochem.,* 72, 248, 1976.
6. Wilkinson, B. S., Fregert, S., Magnusson, B., Vandemann, H-J., Calnan, C. D., Cronin, E., Hjorth, N., Maibach, H. I., Malten, K. E., Menethini, C. L., and Pirilä, V., Terminology of contact dermatitis, *Acta Derm. Venereol.,* 50, 287, 1970.
7. Logan, R. A. and White, I. R., Carba mix is redundant in the patch test series, *Contact Derm.,* 18, 303, 1992.
8. Rademaker, M. and Forsythe, A., Carba mix is a useful indicator of rubber sensitivity, in *Current Topics in Contact Dermatitis,* Frosch, P. J., Dooms-Goossens, A., Lachapelle, J.-M., Rycroft, R. J. G., and Sheper, R. J., Eds., Springer-Verlag, Berlin, 1989, 136.
9. Andersen, K. A., Burrows, D., Cronin, E., Dooms-Goossens, A., Rycroft, R. J. G., and White, I. R., Recommended changes to standard series, *Contact Derm.,* 19, 389, 1988.
10. Beck, M. H., Owen-Smith, V., and Cherry, N., Unpublished results from the Occ-Derm Project.

12

Clinical Testing of Occupation-Related Glove Sensitivity

An E. Dooms-Goossens

TABLE OF CONTENTS

I. Introduction 171
II. The Contact Urticaria Syndrome 172
III. Allergic Contact Eczema 174
IV. Penetration of Chemicals Through Gloves 176
V. Conclusion 177
References 181

I. INTRODUCTION

Glove dermatitis is a frequent finding in certain job-related skin reactions for which some professions are at risk: the building trades, the mechanical industries, cleaning, housekeeping, hairdressing, and particularly the medical and paramedical professions with increasing frequency in recent years (e.g., Fisher[1]). Protective gloves often come in contact with damaged or inflamed skin, which conditions may have been induced by previously encountered contact allergens or, more likely, by irritants undoubtedly associated with these various professions. Hence, the potential glove allergens penetrate more easily into the damaged skin than would otherwise be the case, thus giving rise to sensitization. Moreover, people with a history of atopy, and particularly those suffering from hand eczema, are liable to develop adverse reactions to gloves (e.g., von Hintzenstern et al.[2] and Lammintausta and Kalimo[3]). That gloves can cause intolerance reactions not only in those wearing them became particularly apparent with the contact urticaria syndrome caused by latex, but this may also occur with contact eczema.

In general, allergic reactions due to gloves are either contact eczema reactions (type IV) or the contact urticaria syndrome (type I), and rubber gloves are much more often reported to be causes of contact allergic reactions than are plastic or leather gloves.[4-6]

The allergens responsible for contact eczema to rubber gloves are generally chemicals added during manufacture, such as vulcanizing agents, antioxidants, and accelerators. Other causal agents have also been reported: starch[7] — indeed, chemicals such as epichlorhydrin, sorbic acid, and even isothiazolinone derivatives may be present in such powders;[8] cetyl pyridinium chloride present on a glove's surface;[9] latex protein,[10] although an additive already present in the latex was not completely ruled out as the causal factor;[11] and even a bacterial endotoxin in gamma-radiation-treated gloves[12] and residual ethylene oxide.[13] In plastic gloves, coloring agents or additives also present in rubber gloves have been reported to cause reactions,[4-6,14] while in leather gloves, chromium salts are the main sensitizers.

Contact urticarial reactions are generally due to latex proteins. However, some authors have also considered cornstarch,[15-17] although it is mostly contaminated with latex protein, and certain rubber additives (e.g., Geier and Fuchs,[18] Wrangsjö et al.,[19] and Heese et al.[8]) to be the causal agents in the contact urticaria syndrome. Moreover, protein contact dermatitis to latex has also been described (e.g., Kleinhans[20]).

Glove-related dermatitis can also be caused by various chemicals penetrating through the gloves.[21] Examples are nickel,[21,22] nitrogen mustard,[23] glycerylmonothioglycolate,[24] epoxy resins,[25] acrylic monomers (e.g., Pegum and Medhurst[26] and Afsahi et al.[27]), and nitroglycerin.[28]

II. THE CONTACT URTICARIA SYNDROME

Over a 6-year period from December 1986 to December 1992, 46 patients (mean age: 31 years old) were diagnosed in our clinic as suffering from contact urticarial reactions to latex gloves (Table 1). The overwhelming majority of them were women (42) and presented antecedents of atopy: 25 personal and familial, 7 only personal, and 5 only familial. This concurs with what other authors have reported (e.g., Turjanmaa[29]). The other affected persons suffered from hand eczema (allergic and/or irritant).[30,31]

All of the cases were job related: 25 out of the 46 were medical or paramedical personnel (13 nurses, 6 laboratory technicians, 3 physical therapists, 2 dentists, 1 physician), 18 were people who (sometimes also in addition to their jobs) did household work, 3 who came in contact with food (including 1 butcher, 1 baker, and 1 cook), 2 were cleaners, and 2 were hairdressers. There was also an engineer in a latex glove manufacturing plant.

In most cases, the urticaria occurred on the back of the hands and the fingers, sometimes on the wrists, and occasionally on the face, particularly the eyelids. In one case, the eyelids were the only area affected. Such reactions are mainly caused by airborne contact with glove powder[15,16] which carries the allergens, but sometimes the reaction is due to transfer of the allergen by the

TABLE 1 The Contact Urticaria Syndrome to Latex (1986–1992)

No.	Sex	Age	Occupation	Atopy	Total IgE (U/ml)	IgE latex
1	F	27	Hairdresser	+	126	+
2	F	30	Kinesthesist	+	ND	+
3	F	34	Nurse	−	44.6	−
4	F	30	Nurse	+	84.3	++
5	F	27	Housekeeper	+	287	−
6	M	36	Engineer	−	176	+
7	F	22	Laboratory technician	+	5338	ND
8	F	37	Housekeeper	+	ND	±
9	F	29	Nurse	+	1744	++
10	F	31	Housekeeper	+	188	+
11	F	30	Nurse/housekeeper	+	100	−
12	F	36	Housekeeper	+	354	++
13	F	35	Housekeeper	+	1886	++
14	F	33	Housekeeper	+	ND	+
15	F	31	Butcher/housekeeper	+	2456	+++
16	F	38	Housekeeper	+	1139	−
17	F	29	Housekeeper	+	3063	++
18	F	29	Housekeeper	+	4385	++
19	F	27	Cleaner	+	ND	ND
20	F	23	Beautician/housekeeper	+	1892	++
21	F	33	Cook/housekeeper	+	268	ND
22	F	27	Baker/housekeeper	−	398	+
23	F	22	Laboratory technician	−	77.6	−
24	F	28	Nurse	+	79.5	−
25	F	26	Nurse	+	1602	−
26	F	42	Housekeeper	+	ND	ND
27	M	83	Housekeeper	−	191	+
28	F	31	Nurse	+	ND	ND
29	F	44	Farmer	+	ND	ND
30	F	30	Nurse	+	200	++
31	F	23	Dentist	+	404	±
32	F	24	Nurse	−	ND	ND
33	F	23	Kinesthesist	+	394	−
34	M	26	Kinesthesist	+	205	−
35	F	22	Dentist	+	ND	ND
36	F	27	Nurse	+	ND	ND
37	F	25	Housekeeper	+	ND	ND
38	F	30	Laboratory technician	−	ND	ND
39	F	28	Laboratory technician	+	ND	ND
40	F	47	Cleaner	+	ND	ND
41	F	84	Nurse	+	ND	ND
42	F	30	Laboratory technician	−	43.7	±
43	M	30	Housekeeper	+	244	+
44	F	25	Laboratory technician	−	207	±
45	F	23	Nurse	+	ND	−
46	F	27	Physician	+	229	+

gloved hand or by the hand after the glove is removed. In a few instances, extracutaneous symptoms (rhinitis, conjunctivitis, asthma) were also present, and some of the patients reported having occasionally experienced severe intolerance reactions when they were touched by paramedical personnel wearing gloves.[32-35] In six patients, associations were seen with a fruit allergy, i.e., chestnuts, bananas,[36-38] and/or kiwi fruit.

Of the 46 patients with the contact urticaria syndrome to latex, 21 also presented positive patch test reactions to contact allergens, the most frequent of which was nickel (9 patients). Among the rubber allergens, thiuram mix gave a positive reaction in 2 patients (in 1 case associated with carba mix) as did paraphenylenediamine and diaminodiphenylmethane (both *p*-aminobenzene compounds) in 2 other patients and colophony in another patient. The combined contact urticaria syndrome to latex and allergic contact dermatitis to rubber additives has been reported.[39,40]

Most of the patients diagnosed as having a latex sensitivity were administered prick tests with latex glove extracts, which seems to be the most appropriate method.[41] In a few instances, however, the prick test with the latex was negative while a scratch and/or use test with the responsible glove was positive. Specific IgEs for latex were sought in 32 cases (see Table 1), 18 of which gave a positive (+, ++, or +++) and 4 a weak positive (±) result (69%). This concurs with the literature on the subject.[31,41,42] Generally, the patients with high total IgE levels presented with the highest sensitivity to latex.[32]

III. ALLERGIC CONTACT ECZEMA

From January 1978 through December 1992, 9170 patients were examined in our Contact Allergy Unit. The MOAHL-index percentages are

Males:	34%
Occupational dermatitis:	23%
Atopy:	27%
Hand dermatitis:	43%
Leg ulcers/stasis dermatitis:	7%

Of them, 4620 (52.5%) showed at least one positive patch test reaction and 155 (1.7% of the total patient population tested) presented with a clear glove-related, occupation-induced dermatitis. This number includes two laboratory technicians who compounded paints, one of whom was epoxy-resin sensitive and the other colophony-sensitive. They both reacted only to the rubber gloves they had worn and that had been contaminated with these allergens.

The mean age of these patients was 38 years old; 111 (72%) were female and 44 (28%) were male. The dermatitis was due to rubber gloves in 149 cases, to leather gloves in 3 cases (in 1 of these cases there was also a rubber-glove sensitivity), and to gloves made of PVC in 1 case, of nitrile rubber in

TABLE 2 The "Glove" Allergens Found in 153 Cases of Occupation-Related Contact Dermatitis

1.	Thiuram mix	87
2.	Carba mix	47
3.	Paraphenylenediamine (PPD)	28
4.	Diaminodiphenylmethane	25
5.	Colophony	14[a]
6.	Mercapto mix	11
	PPD-black rubber mix	11
7.	p-Amino azobenzene	7
8.	Disperse orange 3	6
9.	Diphenylguanidine	5
10.	Dihydroxydiphenyl	2
	Disperse yellow 3	2
	Dithiomorpholine	2
	Diphenylthiourea	2
	Potassium dichromate	2
11.	Hydroquinone monobenzylether	1
	Trimethyldihydroquinone	1
	Dimethylthiourea	1
	Ethylenethiourea	1
	Diethylthiourea	1
	Phenyl β-naphthylamine	1
	Morpholinyl-mercaptobenzothiazole	1
	Cobalt	1

[a] In six other cases the colophony allergy could also account for adhesive-tape allergy.

1 case, and of polyethylene in 1 case (the last patient also reacted to rubber gloves).

The nature and the frequency of the glove allergens found for this patient population are given in Table 2. The occupations concerned are listed in Table 3.

Of the 3 patients (1 gardener, 1 chauffeur, 1 worker in the construction industry) who wore leather gloves, 2 reacted to potassium dichromate and 1 to PPD, diaminodiphenylmethane, and azo dyes (p-aminoazobenzene, disperse orange 3, disperse yellow 3), which we considered to be associated with the yellow color of his gloves. For the 19 patients who reacted to their own gloves, the data are given in Table 4.

Thiuram mix is the most frequent positive rubber mix allergen observed in a patient population suffering from occupation-related glove allergy and is most often related to housework and the medical and paramedical professions. This is in agreement with the results of other studies (e.g., von Hintzenstern et al.[2] and Lammintausta and Kalimo[3]). Although thiuram mix and carba mix are most often associated,[43] the latter was found without thiuram sensitivity in 12 cases.[44]

Furthermore, there were also three patients who had presented eczematous reactions upon contact with rubber gloves worn by medical professionals. Two

TABLE 3 The Occupations Involved in Glove Dermatitis (N = 155)

1.	Housework	57
2.	Nursing	16
3.	Metalworking	14
4.	Gardening/vegetable growing	13
5.	Cleaning	12
6.	General practice medicine (physicians)	8
	Construction	8
7.	Laboratory technology	7
	Farming (cattle)	7
8.	Hairdressing	6
	Transport (chauffeur)	6
9.	Cooking/restaurant	5
10.	Plastics industry	4
11.	Meat handling	3
12.	Backing	2
	Mining	2
13.	Fish sales	1
14.	Sports (soccer goalkeeper)	1

of them, one with an amputation wound on his leg and the other with a stoma, reacted to thiuram mix, while the third patient's lips and gingivae became red and swollen after he had been to the dentist and reacted strongly to colophony, which, according to the dentist, had not been used in the dental procedure. Similar cases have been reported by Goh.[45]

IV. PENETRATION OF CHEMICALS THROUGH GLOVES

The chemicals that we found to be causes of dermatitis in relation to their penetration through gloves were acrylate compounds in dental workers. Figure 1 illustrates the typical dermatitis present on the fingertips of a dentist allergic to UV-curing acrylate compounds — triethyleneglycol dimethacrylate, ethylene glycol dimethacrylate, and butanedioldimethacrylate — even though he always wore gloves when handling these compounds. We also observed allergic reactions in printers who came in contact with ink or flexible printing plates containing UV-curing acrylates. Furthermore, epoxy resins in mechanics in the aircraft industry, thioglycolates in hairdressers, and acrylamide in a laboratory technician[46] were found to penetrate gloves and cause allergic contact dermatitis. Figure 2 shows contact dermatitis lesions that this last patient presented on the back of the hand and that were caused by sensitivity to acrylamide used in gel electrophoresis. She always wore latex protective gloves. With this laboratory technician, we obtained positive patch test reactions when pieces of latex (Figure 3) as well as PVC gloves (Figure 4) painted with a 30% solution of the chemical on the opposite side were applied to the skin for 2 days.

TABLE 4 Patch Test Results Obtained in Patients Who Reacted to Their Own Gloves

1.	Construction worker	Rubber glove, colophony
2.	Metalworker	Rubber glove
3.	Plastics (paint) industry worker	Only colophony contaminated rubber gloves, colophony (plastic series negative)
4.	Plastics industry worker	(Red) rubber glove, PPD, diaminodiphenylmethane, p-aminoazobenzene, disperse orange 3, (plastic series negative)
5.	Nurse	Rubber glove, polyethylene glove, formaldehyde (plastic series not tested)
6.	Housekeeper	Rubber glove, thiuram mix
7.	Cleaner	Blue rubber glove, cobalt
8.	Hairdresser (hobby), fish handler	Rubber glove (plastic series negative)
9.	Metalworker	Nitrile glove
10.	Laboratory technician (chemistry)	PVC glove, (plastic series negative)
11.	Gardener	Rubber glove, PPD mix
12.	Laboratory technician (chemistry)	Rubber glove, dihydroxydiphenyl, diphenylthiourea
13.	Laboratory technician (biology)	Rubber glove, thiuram mix
14.	Physician	Rubber glove, thiuram mix
15.	Metalworker	Black rubber glove, thiuram mix
16.	Laboratory technician (chemistry)	Black rubber glove
17.	Gardener/household work	Rubber glove, thiuram mix, carba mix
18.	Plastic (paints) industry worker	Only epoxy-resin contaminated glove, epoxy resin
19.	Nurse	Rubber gloves, thiuram mix, carba mix, dithiodimorpholine

Note: A rubber series was tested only in patients 12–14, 17, and 19.

Recently, we also observed a patient who compounded polyurethane and silicone resins and had developed a contact allergy to the epoxy silane compound.[58] This compound penetrated through his gloves, both those made of polyethylene and those made of PVC (PVC support jersey).

V. CONCLUSION

Analyses of our data on glove-related occupational dermatitis problems show concordance with the literature data on the contact urticaria syndrome, on allergic contact eczema, as well as on penetration of chemicals through gloves.

Obviously, an important number of glove-related occupationally induced cases of contact urticaria and contact eczema could be substantially reduced if the presence of important allergens could be reduced or eliminated. This is

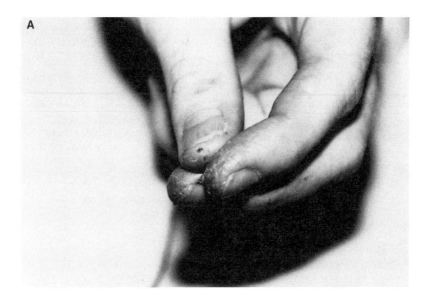

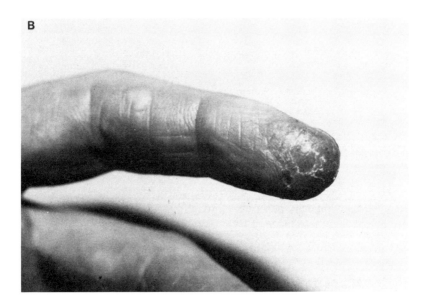

FIGURE 1 Typical dermatitis present on the fingertips of a dentist allergic to UV-curing acrylate compounds. This patient always wore latex gloves when handling these materials.

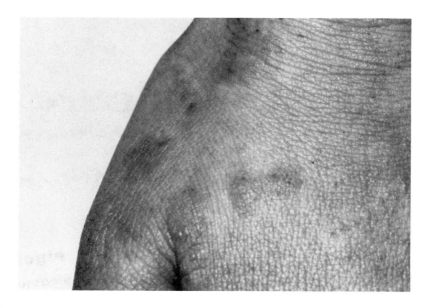

FIGURE 2 Allergic contact dermatitis on the back of the hands due to acrylamide that passed through latex gloves worn by a lab technician performing gel electrophoresis.

particularly important as the patients with glove dermatitis are generally young. For proteins, washing the gloves after molding, before they are dried, and steam sterilization, may decrease or even suppress immediate-type allergenicity.[47] With regard to additives such as thiuram derivatives and carbamates, the release of which can be measured, their presence should also be reduced to a minimum and, perhaps, even be restricted by law.[48] For people allergic to glove materials, allergen alternatives are available (e.g., Heese et al.[8] and Rich et al.[49]) although apparently the labels placed on gloves cannot always be relied upon.[48]

When chemicals penetrate the gloves, the risk of sensitization, of course, is increased because of the occlusion effect. This belies the sense of security patients feel when they wear "protective" gloves. Hence, information about penetration rates and the development of suitably safe products are necessary for people who work with hazardous chemicals. Therefore, the risks should be analyzed, and the appropriate gloves should be selected.[50,51] Different glove materials have already been shown to provide varying degrees of protection to specific irritants and allergens,[52-55] and data have also been compiled in a database on the subject.[56] Moreover, it is heartening to see glove manufacturers making efforts to respond to specific requirements.[57]

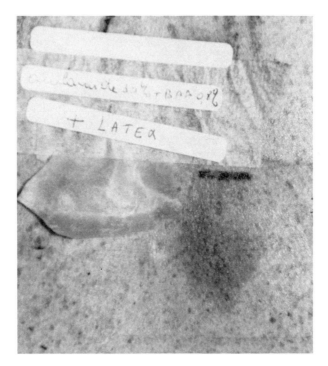

FIGURE 3 Positive test in the patient shown in Figure 2 to a solution containing 30% acrylamide and 0.8% N,N'-methylenebisacrylamide passing through a piece of latex.

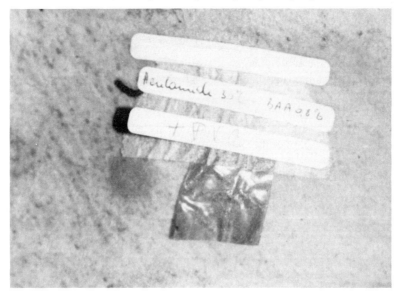

FIGURE 4 Positive test in the patient shown in Figure 2 to a solution containing 30% acrylamide and 0.8% N,N'-methylenebisacrylamide passing through a piece of PVC glove.

REFERENCES

1. Fisher, A. A., Management of allergic contact dermatitis due to rubber gloves, *Cutis*, 47, 301, 1991.
2. von Hintzenstern, J., Heese, A., Koch, H. U., Peters, K. P., and Hornstein, O. P., Frequency, spectrum and occupational relevance of type 4 allergens to rubber chemicals, *Contact Derm.*, 24, 244, 1991.
3. Lammintausta, K. and Kalimo, K., Sensitivity to rubber. Study with rubber mixes and individual rubber chemicals, *Dermatosen*, 33, 204, 1985.
4. Estlander, T., Jolanki, R., and Kanerva, L., Dermatitis and urticaria from rubber and plastic gloves, *Contact Derm.*, 14, 20, 1986.
5. Frosch, P. J., Born, C. M., and Schütz, R., Kontaktallergien auf Gummi-, Operations-, unde Vinylhandschuhe, *Hautarzt*, 38, 210, 1987.
6. Guillet, M. H., Ménard, N., and Guillet, G., Sensibilisation de contact aux gants en vinyl. A propos d'un cas de polysensibilisation aux gants médicaux, *Ann. Dermatol. Vénéreol.*, 118, 723, 1991.
7. Milkovic-Kraus, S., Glove powder as a contact allergen, *Contact Derm.*, 26, 198, 1992.
8. Heese, A., von Hintzenstern, J., Peters, K.-P., Koch, H. U., and Hornstein, O. P., Allergic and irritant reactions to rubber gloves in medical health services: spectrum, diagnostic approach, and therapy, *J. Am. Acad. Dermatol.*, 25, 831, 1991.
9. Castelain, M. and Castelain, P.-Y., Allergic contact dermatitis from cetyl pyridinium chloride in latex gloves, *Contact Derm.*, 28, 118, 1993.
10. Wyss, M., Elsner, P., Wüthrich, B., and Burg, G., Allergic contact dermatitis to natural latex, *J. Am. Acad. Dermatol.*, 27(4), 650, 1992.
11. Heese, A., Reply, *J. Am. Acad. Dermatol.*, 27(4) 651, 1992.
12. Shmunes, E. and Darby, T., Contact dermatitis due to endotoxin in irradiated latex gloves, *Contact Derm.*, 10, 240, 1984.
13. Royce, A. and Moore, K. S., Occupational dermatitis caused by ethylene oxide, *Br. J. Ind. Med.*, 12, 169, 1955.
14. Jolanki, R., Kanerva, L., and Estlander, T., Organic pigments in plastic can cause allergic contact dermatitis, *Acta Derm. Venereol.*, Suppl. 134, 95, 1987.
15. van der Meeren, H. L. M. and van Erp, P. E. J., Life-threatening contact urticaria from glove powder, *Contact Derm.*, 14, 190, 1986.
16. Fisher, A. A., Contact urticaria due to cornstarch surgical glove powder, *Cutis*, Nov., 307, 1986.
17. Assalve, D., Cicioni, C., Perno, P., and Lisi, P., Contact urticaria and anaphylactoid reaction from cornstarch surgical glove powder, *Contact Derm.*, 19, 61, 1988.
18. Geier, J. and Fuchs, T., Kontakturtikaria durch Gummihandschuhe, *Z. Hautkr.*, 65(3), 267, 1990.
19. Wrangsjö, K., Mellström, G., and Axelsson, G., Discomfort from rubber gloves indicating contact urticaria, *Contact Derm.*, 15, 79, 1986.
20. Kleinhans, D., Sofort-typ Allergie gegen Latex: Kontakt-Urticaria und Ekzem, *Akt. Dermatol.*, 10, 227, 1984.
21. Moursiden, H. T. and Faber, O., Penetration of protective gloves by allergens and irritants, *Trans. St. John's Hosp. Dermatol. Soc.*, 59, 230, 1973.

22. Wall, L. M., Nickel penetration through rubber gloves, *Contact Derm.*, 6, 461, 1980.
23. Thomsen, K. and Mikkelsen, H. I., Protective capacity of gloves used for handling of nitrogen mustard, *Contact Derm.*, 1, 268, 1975.
24. Storrs, F., Permanent wave contact dermatitis: contact allergy to glyceryl monothioglycolate, *J. Am. Acad. Dermatol.*, 11, 74, 1984.
25. Pegum, J. S., Penetration of protective gloves by epoxy resins, *Contact Derm.*, 5, 281, 1979.
26. Pegum, J. S. and Medhurst, F. A., Contact dermatitis from penetration of rubber gloves by acrylic monomer, *Br. Med. J.*, 2, 141, 1971.
27. Afsahi, S. P., Sydiskis, R. J., and Davidson, W. M., Protection by latex or vinyl gloves against cytotoxicity of direct bonding adhesives, *Am. J. Orthod. Dentofacial Orthop.*, 93(1), 47, 1988.
28. Hogstedt, C. and Ståhl, R., Skin absorption and protective gloves in dynamite work, *Am. Ind. Hyg. Assoc. J.*, 41, 367, 1980.
29. Turjanmaa, K., Incidence of immediate allergy to latex gloves in hospital personnel, *Contact Derm.*, 17, 270, 1987.
30. Taylor, J. S., Cassettari, I., Wagner, W., and Helm, T., Contact urticaria and anaphylaxis to natural latex, *J. Am. Acad. Dermatol.*, 21, 874, 1989.
31. Wrangsjö, K., Wahlberg, J. E., and Axelsson, I. G. K., IgE-mediated allergy to natural rubber in 30 patients with contact urticaria, *Contact Derm.*, 19, 264, 1988.
32. Pecquet, C., Leynadier, F., and Dry, J., Contact urticaria and anaphylaxis to natural latex, *J. Am. Acad. Dermatol.*, 22, 631, 1990.
33. Fisher, A. A., Iatrogenic (intraoperative) rubber glove allergy and anaphylaxis. I, *Cutis*, 49, 17, 1992.
34. Morren, M., Mariën, K., and Dooms-Goossens, A., Latex allergie: zeldzame oorzaak van anafylactische reacties gedurende heelkundige ingrepen of na gynaecologisch of stomatologisch onderzoek, *T. Geneesk.*, 46(9), 683, 1990.
35. Leynadier, F., Pecquet, C., and Dry, J., Anaphylaxis to latex during surgery, *Anaesthesia*, 44, 547, 1989.
36. Turjanmaa, K., Räsänen, L., Lehto, M. I., Mäkinen-Kiljunen, S., and Reunala, T., Basophil histamine release and lymphocyte proliferation tests in latex contact urticaria, *Allergy*, 44, 181, 1989.
37. M'Raihi, L. F., Charpin, D., Pons, A., Bongrand, P., and Vervloet, D., et al., Cross-reactivity between latex and banana, *J. Allergy Clin. Immunol.*, 87, 129, 1991.
38. Ceuppens, J. L., Van Durme, P., and Dooms-Goossens, A., Severe food-induced anaphylaxis in patients with latex allergy, *Lancet*, 339, 493, 1992.
39. Turjanmaa, K. and Reunala, T., Latex-contact urticaria associated with delayed allergy to rubber chemicals, in *Current Topics in Contact Dermatitis*, Frosch, P. J., Dooms-Goossens, A., Lachapelle, J.-M., Rycroft, R. J. G., and Scheper, R. J., Eds., Springer-Verlag, Berlin, 1989, 460.
40. Görtz, J. and Goos, M., Immediate and late-type allergy to latex: contact urticaria, asthma and contact dermatitis, in *Current Topics in Contact Dermatitis*, Frosch, P. J., Dooms-Goossens, A., Lachapelle, J.-M., Rycroft, R. J. G., and Scheper, R. J., Eds., Springer-Verlag, Berlin, 1989, 457.

41. Turjanmaa, K., Reunala, T., and Räsänen, L., Comparison of diagnostic methods in latex surgical glove contact urticaria, *Contact Derm.,* 19, 241, 1988.
42. Jaeger, D., Kleinhans, D., Czuppon, A. B., and Baur, X., Latex-specific proteins causing immediate-type cutaneous, nasal, bronchial and systemic reactions, *J. Allergy Clin. Immunol.,* 89(3), 759, 1992.
43. Van Ketel, W. G. and Van den Berg, W. H. H. W., The problem of the sensitisation to dithiocarbamates in thiuram-allergic patients, *Dermatologica,* 169, 70, 1984.
44. Rademaker, M. and Forsyth, A., Carba mix: a useful indicator of rubber sensitivity, in *Current Topics in Contact Dermatitis,* Frosch, P. J., Dooms-Goossens, A., Lachapelle, J.-M., Rycroft, R. J. G., and Scheper, R. J., Eds., Springer-Verlag, Berlin, 1989, 136.
45. Goh, C. L., Contact allergy to surgeon's gloves in their patients, *Contact Derm.,* 20, 223, 1989.
46. Dooms-Goossens, A., Garmyn, M., and Degreef, H., Contact allergy to acrylamide, *Contact Derm.,* 24, 70, 1991.
47. Leynadier, F., Tran Xuan, T., and Dry, J., Allergenicity suppression in natural latex surgical glove, *Allergy,* 46, 619, 1991.
48. Knudsen, B. B., Larsen, E., Egsgaard, H., and Menné, T., Release of thiuram and carbamates from rubber gloves, *Contact Derm.,* 28, 63, 1993.
49. Rich, P., Belozer, M. L., Norris, P., and Storrs, F. J., Allergic contact dermatitis to two antioxidants in latex gloves: 4,4′-thiobis (6-*tert*-butyl-*meta*-cresol) — (Lowinox 44S36) and butylhydroxyanisole. Allergen alternatives for glove-allergic patients, *J. Am. Acad. Dermatol.,* 24, 37, 1991.
50. Berardinelli, S. P., Prevention of occupational skin disease through use of chemical protective gloves, *Dermatol. Clin.,* 6, 115, 1988.
51. Leinster, P., Bonsall, J. L., Evans, M. J., and Lewis, S. J., The application of test data in the selection and use of gloves against chemicals, *Ann. Occup. Hyg.,* 34(1), 85, 1990.
52. Chéron, J., Guenier, J.-P., and Moncelon, B., Tableaux récapitulatifs, *Travail et Sécurité,* Ed., I.N.R.S., No. 573 (order code 8131), 1976.
53. Blanken, R., Nater, J. P., and Veenhoff, E., Protection against epoxy resins with glove materials, *Contact Derm.,* 16(1), 46, 1987.
54. Darre, E., Vedel, P., and Jensen, J. S., Skin protection against methylmethacrylate, *Acta Orthop. Scand.,* 58(3), 236, 1987.
55. Dinter-Heidorn, H. and Carstens, G., Comparative study on protective gloves for handling cytotoxic medicines: a model study with carmustine, *Pharm. Weekbl. [Sci.],* 14, 180, 1992.
56. Mellström, G., Protective effect of gloves — compiled in a database, *Contact Derm.,* 13, 162, 1985.
57. Roed-Petersen, J., A new glove material protective against epoxy and acrylate monomer, in *Current Topics in Contact Dermatitis,* Frosch, P. J., Dooms-Goossens, A., Lachapelle, J.-M., Rycroft, R. J. G., and Scheper, R. J., Eds., Springer-Verlag, Berlin, 1989, 603.
58. Dooms-Goossens, A., unpublished data, 1993.

13

Allergologic Evaluation and Data on 173 Glove-Allergic Patients

Angelika Heese, Klaus-Peter Peters,
Hans Uwe Koch, and Otto Paul Hornstein

TABLE OF CONTENTS

I. Introduction 186
II. Methods 186
    A. Evaluation of Contact Dermatitis 186
    B. Evaluation of Contact Urticaria 187
        1. Prick and Scratch Tests 187
        2. Latex RAST® 189
    C. Evaluation of Protein Contact Dermatitis 189
    D. Additional Tests 189
        1. Glove Use Tests 189
        2. Total Serum IgE 189
III. Results 190
    A. Main Aspects of 94 Patients With Delayed-Type Allergies 190
        1. Personal History 190
        2. Results of Patch Tests 190
        3. Trends in Prevalence 193
    B. Main Aspects in 104 Patients With Immediate-Type Allergies 193
        1. Personal History 193
        2. Results of Prick, Scratch, and Scratch-Chamber Tests 195
        3. Increase in Latex Allergy 196
    C. Results of Laboratory Tests 196
        1. Latex and Banana RAST® 196
        2. Total Serum IgE 196
IV. Discussion 196
    A. Test Procedure 197
        1. Type IV Allergens 197
        2. Type I Allergens 198
        3. Latex RAST® 198

B. Unusual Allergens 199
     1. Type IV Allergens 199
     2. Type I Allergens 199
  C. Current Trends and Possible Risk Factors 200
V. Conclusion 202
References 202

I. INTRODUCTION

Rubber gloves are indispensable in surgical procedures. Moreover, they protect wet-workers against skin damage and employees in the medical health services as well as their patients against microbial contamination. With the rising public awareness of acquired immune deficiency syndrome (AIDS), the use of rubber gloves in medicine has increased, along with an increase in allergic reactions to their ingredients.[1-9] Aside from irritant dermatitis due to the effects of glove occlusion, delayed-type allergies to accelerators, mainly of the thiuram group, and immediate-type allergies to a panel of water-soluble proteins account for most glove-related skin symptoms.[1-7,10-28]

The purpose of our retrospective study which included data from January 1989 to December 1992 was to analyze 173 glove-allergic patients, with the main emphasis on allergens and prevalence rates of immediate- and delayed-type reactions, including present trends and possible risk factors. In addition, the sensitivity of a commercially available latex RAST® (radioallergosorbent test)* was evaluated comparing the results of this *in vitro* method with both prick test reactions to latex fluid and the personal history of a series of latex-allergic patients.

II. METHODS

From January 1989 to December 1992, 173 patients with allergies to rubber gloves, 119 (68.8%) females and 54 (31.2%) males, were investigated at the Department of Dermatology, University of Erlangen-Nuremberg. Depending on each individual's detailed personal and occupational history as well as clinical symptoms, the allergologic evaluation and *in vitro* analyses were performed as shown in Figure 1.

A. Evaluation of Contact Dermatitis

In all patients suffering from glove-induced hand eczema, patch tests were carried out with the standard series of the International Contact Dermatitis Research Group (ICDRG)[14] supplemented by the 3% carba mix (ZDC, ZDBC, DPG), 2% carba mix (ZDC, ZDBC), and Euxyl K400®**. Additional patch tests comprised a battery of 26 rubber chemicals (Hermal, Reinbek, Germany,

---
 * Registered trademark of Pharmacia, Uppsala, Sweden.
**Registered trademark of Schülke and Mayr GmbH, Norderstedt, Germany.

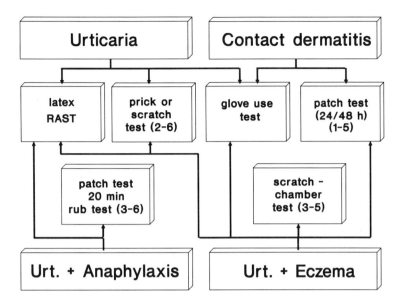

FIGURE 1 Diagnostic evaluation of glove-induced allergies; (1) standard series (see text), (2) rubber chemicals (Table 1), (3) low-ammoniated, accelerator-free latex fluid, (4) piece of the causative glove, (5) glove powder, (6) watery glove extract.

Table 1), material from the patient's glove, accelerator-free latex fluid, and glove powder (Biosorb®*, and a glove powder from Paul Hartmann AG, Heidenheim, Germany). If indicated, occupationally relevant substances including glove-permeating chemicals (nickel, p-phenylenediamine, thioglycolates, methyl methacrylates, epoxy resins, etc.)[5,29-33] completed the test program.

Allergens were applied on Finn Chambers®** fixed with Scanpor® tape***. The exposure time comprised 1 day; readings were done after 24, 48, and 72 h and interpreted according to ICDRG recommendations.[14] Patients with delayed-type allergy to glove powder were further evaluated by patch tests with a battery of antimicrobials and preservatives (Hermal, Reinbek, Germany) as well as epichlorhydrin as possible glove powder constituents.

B. Evaluation of Contact Urticaria

1. Prick and Scratch Tests

Depending on the severity of clinical symptoms, the allergologic evaluations in our department were performed as shown in Figure 1. Patients with a glove-induced localized urticaria (stage I)[34] have been investigated by prick tests with

---

* Registered trademark of Johnson & Johnson, Norderstedt, Germany
** Registered trademark of Epitest, Tuusula, Finland
*** Registered trademark of Norgesplaster A/S, Norway

TABLE 1 Battery of Rubber Chemicals Used For Patch Test

Rubber chemicals	Abbreviation	In petrolatum (%)
Tetramethylthiuram disulfide	TMTD	0.25
Tetramethylthiuram monosulfide	TMTM	0.25
Tetraethylthiuram disulfide	TETD	0.25
Dipentamethylenethiuram disulfide	PTD	0.25
Tetramethylthiuram disulfide	TMTD	1.00
Tetramethylthiuram monosulfide	TMTM	1.00
Tetraethylthiuram disulfide	TETD	1.00
Dipentamethylenethiuram disulfide	PTD	1.00
N-cylohexyl-2-benzothiazylsulfenamide	CBS	1.00
Dibenzothiazyl disulfide	DBTD	1.00
Morpholinyl mercaptobenzothiazole	MMBT	0.50
Mercaptobenzothiazole	MBT	2.00
N-isopropyl-N'-phenyl-p-phenylenediamine	IPPD	0.10
N,N'-diphenyl-p-phenylenediamine	DPPD	0.25
N,N'-di-β-naphthyl-p-phenylenediamine	DBNPD	1.00
N-phenyl-N'-cyclohexyl-p-phenylenediamine	CPPD	0.25
Phenyl-β-naphthylamine	PBN	1.00
Diethyldithiocarbamate zinc	ZDC	1.00
Dibutyldithiocarbamate zinc	ZDBC	1.00
1,3-Diphenylguanidine	DPG	1.00
N,N'-diphenylthiourea	DPTU	1.00
N,N'-dibutylthiourea	DBTU	1.00
Hydroquinone monobenzylether	HMBE	1.00
4,4'-Dihydroxydiphenyl	DHDP	0.10
Hexamethylenetetramine	HMT	1.00
Benzoyl peroxide	BZP	1.00

low-ammoniated, accelerator-free latex fluid (provided by the London Rubber Company, Mönchengladbach, Germany), a self-prepared watery glove extract (1 g of Peha Soft®glove*, incubated in 8 ml physiological saline for 24 h), and two glove powders (Biosorb® and a glove powder from Paul Hartmann AG, Heidenheim, Germany). Physiological saline and histamine (1.7 mg/ml, Allergopharma, Reinbek, Germany) served as negative and positive controls. Scratch tests with a series of rubber chemicals incorporated in petrolatum (thiurams, dithiocarbamates, benzothiazoles) and with material from the causative glove completed the test procedure. If the personal history revealed a glove-related anaphylactic shock reaction the allergologic evaluation was restricted to a 20-min open patch test with the watery glove extract, latex fluid, a piece of the causative glove, and glove powder.[5] An additional rub test (performed by rubbing substances about ten times on the volar aspect of the forearm)[35] or even prick tests with the aforementioned substances, were carried out in patients with negative results but who required emergency care condi-

---

* Registered trademark of Paul Hartmann AG, Heidenheim, Germany.

tions.[5,36] A wheal-and-flare reaction to glove powder was further investigated by a prick test with sorbic acid as a possible glove powder constituent.[5,37]

The prevalence of latex allergy among patients with atopic diseases was evaluated by analyzing prick test results with accelerator-free latex fluid in 201 atopics. With respect to recent reports of a cross-reactivity between latex and banana,[38-42] additional prick tests with pieces of banana flesh were performed in 15 of 104 latex-allergic patients.

2. Latex RAST®*

Specific IgE against latex was determined in 67 (64.4%) of 104 latex-allergic patients, using a commercially available latex RAST®. Among these were 33 patients with an exclusively localized urticaria, 5 patients with a generalized urticaria, 27 patients with an additional involvement of mucous membranes (such as rhinitis, conjunctivitis, or allergic asthma) as well as 2 patients with a latex-induced anaphylactic shock reaction. Results of the latex RAST® above 0.35 PRU (Pharmacia RAST Unit) were considered positive. A further 42 patients with an immediate-type allergy to latex have been evaluated by RAST® for specific IgE antibodies against both latex and banana.

C. Evaluation of Protein Contact Dermatitis

Clinical features of protein-induced immediate-contact dermatitis (ICD) comprise an IgE-mediated urticaria within 30 min to 6 h after exposure to the causative allergen, which is followed by a secondary eczematous alteration within 6 to 48 h later.[5,43-45] Apart from our test program for both urticaria and contact dermatitis, patients with a suspected ICD have been further evaluated by scratch-chamber tests with accelerator-free latex fluid, a piece of the causative glove, and glove powder (readings at 24, 48, and 72 h).

D. Additional Tests

1. Glove Use Tests

The main indications for glove use tests were equivocal or unsatisfactory prick/patch test results not in line with our clinical diagnosis.[5] These tests were performed by instructing the patient to wear gloves 2 h daily on three consecutive days if delayed-type allergy was suspected. In cases of questionable immediate-type allergy, glove use tests were limited to 30 min, however.

2. Total Serum IgE

IgE antibodies were determined in all 173 glove-allergic patients using a commercially available enzyme-linked immunosorbent assay (Synelisa, Elias, Freiburg, Germany).

---

* Registered trademark of Pharmacia, Uppsala, Sweden.

III. RESULTS

The evaluation of our 173 glove-allergic patients revealed a delayed-type allergy to glove ingredients in 94 (54.3%) and an immediate-type allergy in 104 (60.1%) cases; 25 (14.5%) patients showed a clinically relevant combined delayed- and immediate-type allergy, and thus were considered in both collectives.

A. Main Aspects of 94 Patients With Delayed-Type Allergies

1. Personal History

The mean age of 94 patients with delayed-type allergies to rubber gloves was 38.8 ± 11.3 years (range: 19.7 to 69.5) including 62 females and 32 males with a mean age of 37.5 and 41.0 years, respectively. The mean daily contact to gloves lasted 1.5 ± 1.8 h (range 1 to 6). Clinical manifestations of glove intolerances occurred about 8 weeks to 30 years (mean: 13.5 ± 6.3 years) after having started to wear gloves regularly (Table 2).

Of 94 patients, 32 (34%) had a personal history of atopic diseases mostly comprising allergic rhinitis (78.1%), whereas an additional 12 (12.8%) proved to have positive prick tests to a battery of 22 common inhalant allergens (Allergopharma, Reinbek, Germany) without any concomitant clinical manifestation, and 34 (36.2%) suffered from preexisting hand eczema (Table 2).

The main occupations among the 94 patients were the medical professions in 34 (36.2%) and housekeepers/employees in gastronomy in 30 (31.9%) cases. Other professions were less frequently represented (1.1 to 7.4%) (Tables 2 and 3).

2. Results of Patch Tests

The allergologic evaluation of 94 patients with type IV allergies to rubber gloves revealed positive reactions to thiuram mix (TM) in 77.7% (73 patients),

TABLE 2 Differences in the Personal History of Patients With Type I and Type IV Allergies to Rubber Gloves

	Type I allergy (n = 104)	Type IV allergy (n = 94)
Sex ratio (f/m)	2.5	1.9
Mean age (years ± SD)	29.8 ± 9.8	38.8 ± 11.3
Mean daily contact to rubber gloves (hours ± SD)	2.5 ± 2.0	1.5 ± 1.8
Mean asymptomatic interval since initial glove use (years ± SD)	3.2 ± 1.3	13.5 ± 6.3
Preexisting hand eczema	63 (60.6%)	34 (36.2%)
Atopic disease	66 (63.5%)	32 (34.0%)
Medical profession	77 (74.0%)	34 (36.2%)
Housekeeper and gastronomy	12 (11.5%)	30 (31.9%)

TABLE 3 Occupations of Glove-Allergic Patients

	Type I (n = 104)		Type IV (n = 94)	
	Number	%	Number	%
Medical profession	77	74.0	34	36.2
Housekeeper, gastronomy	12	11.5	30	31.9
Biologist	5	4.8	0	0
Electrician, electrical engineer	2	1.9	7	7.4
Child	2	1.9	0	0
Machine fitter, mechanical engineer	1	1.0	5	5.3
Building worker	1	1.0	4	4.3
Chemical worker	1	1.0	3	3.2
Hairdresser	1	1.0	2	2.1
Forester	1	1.0	1	1.1
Offset printer	1	1.0	1	1.1
Photographer	0	0	2	2.1
Driver	0	0	2	2.1
Textile worker	0	0	2	2.1
Postman	0	0	1	1.1

followed by 3% carba mix (CM) in 31.9% (30), 2% carba mix (CM) in 17% (16), mercapto mix (MM) in 11.7% (11), and black-rubber mix (PM) in 7.4% (7). TM was the only positive rubber mix in 43.6% (41 patients), compared to CM and MM each in 3.2% (3), as well as PM in 2.1% (2), respectively. In some patients, positive patch test results to rubber mixes contrasted to negative reactions to their single substances and vice versa. These results, listed in Table 4, were interpreted as false-positive or false-negative.[46] Accordingly, the 3% CM revealed true-positive reactions in 17/30 patients compared to only 10 found with the 2% CM. On the other hand, 13 false-positive patch test results with the 3% CM contrasted with only 6 attributable to the 2% CM (Table 4). The most frequent combination of positive rubber mixes was found to be TM/CM in 26.6% (25 patients).

With respect to the battery of rubber chemicals, tetramethylthiuram disulfide (TMTD) and tetramethylthiuram monosulfide (TMTM) ranked first, and were found in 57 (80.3%) and 53 (74.7%) of 71 TM-allergic patients, respectively. Diethyldithiocarbamate zinc was the most common positive single component in 16 (94.1%) of 17 CM-allergic patients. Morpholinyl mercaptobenzothiazole (MMBT) was the main allergen in 10 (90.9%) of the 11 MM-allergic patients, whereas N-isopropyl-N'-phenyl-p-phenylenediamine (IPPD) was the most frequent single component, positive in all 7 PM-allergic patients.

Aside from patch tests with 0.25% TMTD, TETD, TMTM, and PTD, 48 of the 73 thiuram-allergic patients have further been evaluated with the single components of the TM mix using 1% test concentrations. Notably, 4 (8.3%) of these patients revealed positive patch tests only to the 1% single substances, but reacted negatively with the 0.25% ones. There were no false-positive

TABLE 4 Positive and Negative Patch Test Results With Rubber Mixes and Their Single Components in 94 Patients With Type IV Allergies to Rubber Gloves

	Thiuram mix 1%		Carba mix 3%		Carba mix 2%		Mercapto mix 2%		Black-rubber mix 0.6%	
	Pos.	Neg.	Pos.	Neg.	Pos.	Neg.	Pos.	Neg.	Pos.	Neg.
Single components positive	71	2	17	1	10	8	11	2	7	0
Single components negative	2	19	13	63	6	70	0	81	0	87

reactions to the 1% TETD, TMTD, TMTM, or PTD among 60 TM-negative controls.

Positive patch tests with a piece of the causative glove were found in 86 (91.5%) of our 94 patients. Remarkably, in two individuals with delayed-type allergies to accelerators, two brands of powdered gloves and the originally wrapped glove powder (provided by the glove manufacturer) produced positive patch tests, whereas the same gloves remained negative after having been washed with water. These test results were in line with both patients' history of intolerance reactions, especially to powdered gloves. Further delayed-type allergies could neither be detected in a battery of antimicrobials and preservatives (including sorbic acid) nor to epichlorhydrin as possible glove powder constituents.[5,24,37]

According to single reports in the literature, unpowdered PVC gloves[3,15,47] and accelerator-free latex fluid[48] proved to be additional allergen sources in two and three TM-positive patients, respectively, and thus were responsible for reproducible positive patch test results.

3. Trends in Prevalence

Our retrospective study revealed an overall prevalence of type IV allergies to rubber gloves of 4.2% (94 patients) among all 2220 patients with allergic contact dermatitis who were evaluated during this 4-year interval. Considering 1-year intervals, the prevalence of type IV allergies to rubber gloves was found to have continuously increased from 2.3% (14/613 patients) in 1989 to 3.8% (20/532 patients) in 1990, 5.5% (30/546) in 1991 and 5.7% (30/529) in 1992. The overall prevalence and its yearly increase was higher in females than males (Figures 2 and 3).

B. Main Aspects in 104 Patients With Immediate-Type Allergies

1. Personal History

The mean age of 104 patients with immediate-type allergy to rubber gloves was 29.8 ± 9.8 years (range 5 to 60) including 74 (71.2%) females and 30 (28.8%) males with a mean age of 28.6 and 32.2 years, respectively. The mean daily contact to gloves lasted 2.5 ± 2.0 h (range 0.5 to 8). After having started to wear gloves regularly, an asymptomatic interval of about 6 weeks to 10 years (mean 3.2 ± 1.3 years) could be documented before contact urticaria to gloves occurred (Table 2).

Based on von Krogh and Maibach's classification of the contact urticaria syndrome,[34] and considering additional results of Baur et al.,[49] who succeeded in proving the existence of latex-induced inhalant allergy, the symptoms in our 104 type I-allergic patients were categorized as follows: 60 (57.7%) patients suffered from localized (stage I) and 5 (4.8%) from generalized urticaria (stage II), whereas 34 (32.7%) patients showed an additional involvement of mucous

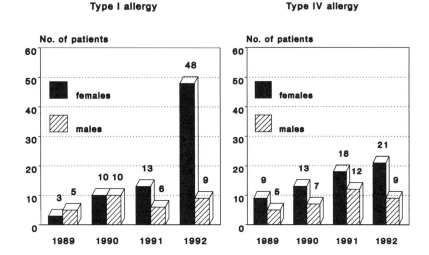

FIGURE 2 Yearly increase of type I and type IV allergies to rubber gloves in Erlangen 1989 to 1992.

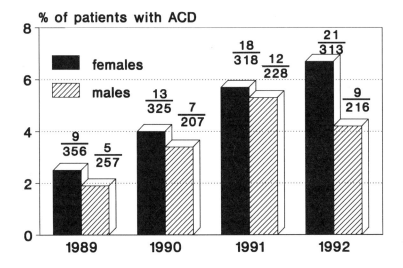

FIGURE 3 Prevalence of type IV allergies to rubber gloves among patients with allergic contact dermatitis (ACD) in Erlangen 1989 to 1992.

membranes, including allergic rhinitis, asthma, or allergic conjunctivitis (stage III or inhalant latex allergy). The history of 5 (4.8%) of the 104 patients revealed a severe intraoperative anaphylactic shock (stage IV) which could be attributed to an immediate-type allergy to latex gloves. Two of these five patients had undergone surgery for malignant brain tumors three times in the past, and two others had experienced previous operative corrections for spina bifida and congenital urological anomalies, respectively.

In 3 of 60 patients with stage I contact urticaria there were additional signs and symptoms of an immediate-contact dermatitis (protein contact dermatitis).[5,43-45] Of the 104 type I-allergic patients, 66 (63.5%) were atopics who mainly suffered from allergic rhinitis (78.8%), conjunctivitis (68.2%) and/or bronchitis (48.5%), and less frequently from flexural eczema (27.3%). In 15 (14.4%) additional patients, positive prick tests to a battery of 22 common inhalant allergens (Allergopharma, Reinbek, Germany) were the only hint for an atopic diathesis. Of the 104 patients, 63 (60.6%) had a history of preexisting hand eczema (Table 2).

Medical professionals ranked first among occupational groups and were found in 77 (74%) of our 104 type I-allergic patients, followed by housekeepers and biologists in 12 (11.5%) and 5 (4.8%) cases, respectively. Other occupational groups were less frequently encountered (Table 3).

2. Results of Prick, Scratch, and Scratch-Chamber Tests

There were 101 of 104 patients with immediate-type allergies to rubber gloves who revealed clear positive prick test results (3+, 4+, according to Ring[50]) to low-ammoniated latex milk at the 20-min reading. In comparison, using the self-prepared glove extract, positive prick tests were less frequently seen in 90 (86.5%) of 104 patients.

Three additional patients with a personal history of glove-related urticarial and eczematous skin symptoms did not come up to prick tests with latex fluid, neither at the 20- nor at the 40-min reading. However, all of them presented an urticarial reaction at the 6-h reading (prick test with sterile physiological saline remained negative) followed by an eczematous alteration within 24 h. The positive results of a scratch-chamber test with latex fluid (eczematous reaction after 24 h with increasing strength up to 72 h) corroborated our clinical diagnosis of latex-induced protein contact dermatitis. Having avoided latex gloves, skin symptoms completely cleared in all three patients within 10 days.

A positive immediate response to both scratch tests with the material from a powdered glove and prick tests with glove powder itself could be detected in two additional latex-allergic patients.

The prevalence of an immediate allergy to latex among 201 patients with atopic diseases was found to be 17.4% (35 patients), comprising 25 (12.4%) cases with a history of glove-induced urticarial reactions.

Of the 104 patients, 8 (7.7%) revealed positive scratch tests to different accelerators such as ZDC, ZDBC, MBT, TETD, TMTD, and TMTM. However, the clinical relevance of these reactions remained uncertain in all but one patient who had a reproducible wheal-and-flare reaction to ZDC; his glove-related skin rash only cleared when he stopped using both latex and ZDC-containing gloves.

There were 7 (46.7%) of 15 latex-allergic patients (including 9 atopics) who were evaluated for an additional immediate-type allergy to banana and revealed positive prick tests to pieces of banana flesh (mean wheal diameter of 5 mm). None of them came up to prick tests with physiological saline. Five of the seven banana-allergic patients suffered from concomitant atopic diseases and provided a personal history of angioedema and/or generalized urticaria provoked by both latex gloves and banana.

3. Increase in Latex Allergy

Between January 1989 and December 1992 104 patients with glove-induced immediate allergy to latex were diagnosed in our department. With respect to 1-year intervals, there was a 7-fold increase in the number of latex-allergic patients — from 8 patients in 1989 to 57 in 1992 (Figure 2). This upward trend could mainly be attributed to an increase in latex-allergic females (from 3 patients in 1989 to 48 patients in 1992), mostly belonging to health care professions.

Remarkably, the ratio of patients with latex-induced localized urticaria (stage I) and those with generalized disease (stage II–IV of the contact urticaria syndrome or inhalant allergy) was much higher in 1989/1990 (25:3) than 1991/1992 (35:41), which may indicate a shift to more severe forms of latex allergy.

C. Results of Laboratory Tests

1. Latex and Banana RAST®

Specific IgE antibodies to latex were detected in 44 (65.7%) of 67 latex-allergic patients. The frequency of positive latex RAST® and its individual levels corresponded to the severity of clinical symptoms, as shown in Table 5; 14 (33.3%) of 42 latex-allergic patients revealed a positive RAST® to both latex and banana (mean banana RAST®: 1.01 PRU, range 0.36 to 3.15).

2. Total Serum IgE

Total serum IgE was elevated (>150 U/ml) in 60 (57.7%) of 104 patients with immediate-type and 28 (29.8%) of 94 with delayed-type allergy to rubber gloves. The 60 latex-allergic patients revealed a mean total IgE of 340 U/ml (range: 160 to 1462 U/ml) compared to 184 U/ml (range 160 to 230 U/ml) in 28 of the type IV-allergic patients.

IV. DISCUSSION

The increasing significance of allergies to rubber gloves has been documented by several authors.[1-9,24,26,51,52] In our present retrospective study covering the years 1989 to 1992, the data of 173 glove-allergic patients have been analyzed in detail to provide information about current trends, main and

TABLE 5 Results of Latex RAST® in 67 Patients With Type I Allergy to Rubber Gloves

	Stage I	Stage II	Stage III	Stage IV
Number of patients	33	5	27	2
Positive patients (>0.35 PRU)	18	3	21	2
Positive patients (%)	54.5	60.0	77.8	100
Mean value (PRU)	3.2	5.1	7.9	52.6
Range (PRU)	0.4–12.4	0.5–10.2	0.7–22.2	42.8–62.3

unusual allergens, possible risk factors, as well as new aspects for the allergologic evaluation.

A. Test Procedure

1. Type IV Allergens

According to our retrospective study, all rubber mixes (TM, MM, PM) except the 2% and 3% CM proved to be reliable in the detection of delayed-type allergies to rubber gloves. True-positive reactions were recorded in 94.7% by TM, in 84.6% by MM, and in 100% by PM. The single components of the individual rubber mixes were not superior in the number of true-positive reactions and thus are not suitable to replace the controversially discussed rubber mixes in the standard series. In contrast, the 3% CM (including ZDC, ZDBC, DPG) revealed false-positive (irritant) reactions in 43.3% (13 of 30 CM-positive patients) and therefore seems to be of limited value for the detection of dithiocarbamate-allergic patients. The 2% CM (including ZDC and ZDBC) was not more convincing: ten true-positive patch test results competed with six false-positive and eight false-negative ones. With respect to corresponding results in the literature[2,14,53-55] and considering the fact that most of the CM-positive patients revealed concomitant delayed-type allergies to thiurams due to a possible cross-reaction[54-56] (83.3% in our series of 30 CM-positive patients), the 3% CM had been withdrawn from the European standard series in 1989, but remained a part of our panel. Otherwise 3 (3.2%) of our 94 glove-allergic patients with clinically relevant positive patch tests to 3% CM exclusively would have been missed. Further reports in the literature support our point of view.[57-59] However, a more reliable marker for dithiocarbamate allergy seems to be necessary. According to our results, 1% ZDC in petrolatum was positive in 94.1% of 17 patients with type IV allergies to 3% CM but did not come up in any of 64 CM-negative ones. Hence, we propose 1% ZDC as the marker for CM-allergic patients, provided our findings will be confirmed in a greater number of patients.

Patch tests with single compounds of TM (TETD, TMTM, TMTD, PTD) using 0.25% (according to 1% TM in the European standard series)[14] vs. 1% test concentrations revealed identical results in all but 4 of 48 patients, who only reacted with the 1% derivatives. Because 80 TM-negative controls did not reveal any false-positive patch test to the 1% TM compounds, these are suggested to be superior for the evaluation of TM-allergic patients and thus are proposed to replace the 0.25% compounds in the series of rubber chemicals.

Patch tests with the patient's glove material may be negative, as found in 8.5% of our 94 patients. However, the glove use test revealed positive results in all of them, and, thus, should always be performed in controversial cases.

2. Type I Allergens

At present, prick test materials for the evaluation of latex allergy and the interpretation of the results are not standardized, impeding a comparison of different investigations.[1,4,5,7,8,22,23,28,60] According to our retrospective study, low-ammoniated accelerator-free latex fluid was better than a self-prepared watery glove extract to diagnose latex allergy, revealing positive results in 100% compared to 86.5% of 104 latex-allergic patients. Assuming a 3-mm wheal to represent a positive result and considering the individual personal history, there was no false-positive reaction in any of the 104 latex-allergic subjects nor in 79 latex-negative controls. Corresponding data reported by Fuchs and Wahl[4] and by Leynadier et al.[41] seem to confirm our results. In addition, significant differences in the allergenicity of different brands of latex gloves, mainly due to various amounts of latex proteins,[27,60] have been reported and thus may influence prick test responses to glove extracts. Therefore, we propose a low-ammoniated latex fluid to be used for the evaluation of latex-allergic patients until further well-standardized test materials will be available.

Readings of prick, scratch-chamber, and glove use tests should be performed after 20 and 40 min as well as after 6, 24, 48, and 72 h, especially if an immediate-contact dermatitis is suspected. Otherwise, late-positive reactions, which have been observed in three of our patients as well as in one other latex-allergic patient are at risk to be overlooked.[5,27]

3. Latex RAST®

Our study revealed an overall 65.7% prevalence of positive latex RAST® (44 of 67 latex-allergic patients). In line with Jaeger et al.[7] and Baur et al.,[61] we found a clear correlation between the severity of clinical symptoms and the rate of positive latex RAST® (54.5% and 76.5% positivity in stage I and stage II-IV/inhalant allergy, respectively) as well as the individual RAST® level. With respect to further reports in the literature, specific IgE antibodies to latex have been documented in 7.7 to 100% of investigated latex-allergic patients,[1,4,7,8,11,37,40,60-62] which is not in opposition to our findings. Lower rates of positive latex RAST® of 7.7%[8] and 30.8%,[1] for instance, were observed in collectives of latex-allergic patients without any systemic reactions or inhalant allergy, compared to higher rates of 89.3%[61] and 100%[11] in groups, including a majority of patients with systemic latex allergy.

Hence, the currently available latex RAST® is of value for the confirmation of a latex allergy, however, according to our and other results, is less sensitive than skin prick tests and therefore is unsuitable as a routine screening method for the detection of high risk patients, e.g., prior to surgery.

B. Unusual Allergens

1. Type IV Allergens

Only 3 of our 73 thiuram-allergic patients proved to have an additional reproducible patch test reaction to accelerator-free latex fluid without any concomitant immediate prick test response to latex, neither at the 20- and 40-min nor at the 6- and 24-h reading. Thus, a latex-induced immediate-contact dermatitis, including a late phase reaction and a type IV allergy to accelerators (thiurams) which may be contained in latex fluid,[5,23,51] seemed to be very unlikely. However, a delayed-type allergy to preservatives (as possible components in latex fluid)[63] or even to latex itself, as has been reported only recently,[48] could explain this unusual reaction.

Delayed-type allergies to glove powder (which could not be attributed to sorbic acid or other preservatives as possible glove powder constituents)[37] as well as to PVC gloves were found in two additional TM-positive patients. Because these rare allergen sources[3,15,47,64] may easily be overlooked, especially in the presence of concomitant type IV reactions to common allergens such as thiurams, these substances should be included in the routine patch test program for the evaluation of glove-related contact dermatitis.

2. Type I Allergens

Several authors proved water-soluble proteins with molecular weights between 10 and 70 kD to be responsible for latex allergy.[4,7,10,12,20,21,40,60,65] In contrast, there are conflicting opinions on the existence of an immediate-type allergy to glove powder,[4,7,66-68] a well known skin irritant.[5] However, two of our latex-allergic patients undoubtedly revealed positive prick test results to glove powder, which was in line with their personal history of intolerance reactions to both powdered latex and PVC gloves. Our preliminary and yet unpublished results of the SDS-gel-electrophoretic analysis of glove powder suggest the presence of at least two distinct proteins, and thus seem to confirm the relevance of glove powder as a contact urticaria-inducing agent. With respect to Seggev et al.[68] and Fisher[67], an immediate-type allergy to glove powder may lead to anaphylactic reactions even if only prick tests are applied.

The pathogenetic mechanism of positive prick tests to different accelerators in eight of our, as well as in other, patients[4,5,69-72] remains to be elucidated. False-positive prick test reactions are known to occur especially in atopics, and thus may have been present in four of our eight patients as well. However, the skin symptoms of another patient without concomitant atopic disease and positive immediate prick test response to latex as well as to ZDC only relieved when he avoided both components. Hence, accelerators have to be considered in the evaluation of glove-induced wheal-and-flare reactions, but by far do not play as important a role as latex itself.

The occurrence of a combined immediate-type allergy to latex and banana[38-42] (as well as to avocado and chestnut)[38,39] in some patients was suspected to be due to a cross-reactivity (common epitopes), and the laboratory results of a latex RAST® inhibition by a banana extract seemed to confirm this opinion.[42] Even though our results (7 of 15 latex-allergic patients with an additional immediate-allergy to banana; 14 of 42 latex-allergic patients with positive banana RAST®) are in line with previous reports, it should be noted that more than 50% of our and other patients were atopics,[1,4,7,8,11,23,26,27,51,52,61] who are at a higher risk to develop concomitant allergies to fruits.[62] In addition, the inhibition of latex RAST® by a banana extract, but not vice versa,[42] may be due to an unspecific mechanism elicited by certain ingredients of the banana (e.g., lectines).

C. Current Trends and Possible Risk Factors

The overall prevalence of delayed-type allergies to rubber gloves was 4.2% among 2220 patients with occupationally and nonoccupationally induced allergic contact dermatitis. Higher overall prevalences of 12.5%[3] or even 19%[73] were found in previous studies on occupational dermatitis exclusively. In agreement with other reports,[2,3,5,13-15,18,24,25,54,59] we confirmed thiurams as the most common and important type IV allergens for glove-allergic patients (77.7% of 94 patients) compared to dithiocarbamates (19.1% true-positive reactions in 94 patients) and benzothiazoles (13.8% of 94 patients).

Remarkably, our study revealed a continuous increase in the yearly prevalence of type IV allergies to rubber gloves, from 2.3% (1989) to 5.7% (1992) which is in line with our previous investigations.[6] This upward trend was also established by a sevenfold increase in the yearly number of glove-induced latex allergies (1989/1992: 8/57 patients). With respect to the medical professions present in 74% of our 104 latex-allergic and in 36.2% of 94 type IV-allergic patients, this increase is presumably due to a more frequent use of rubber gloves in medicine as a prophylactic measure against AIDS infection.[1-9,24,26,51,52] The higher proportion of females compared to males established in our (f/m: 119/54) and other studies[1,3,4,7,8,15,23,26,61] on glove-allergic patients corroborates this finding since females represent the majority of health care professionals.

Notably, the ratio of type I and type IV allergies to rubber gloves, which to our knowledge has not been investigated in a large collective so far, increased between 1989 (8:14) and 1992 (57:30). Although the present data only reflect the experiences of our department, they are backed by an internationally reported worrisome increase in immediate allergies to rubber gloves.[1,4,7,8,9,26] Different mean asymptomatic intervals after the initial glove exposure of 3.2 ± 1.3 years in our 104 latex-allergic patients compared to 13.5 ± 6.3 years in our 94 type IV-allergic patients may, in part, be responsible for this overproportional increase of type I allergies to rubber gloves during the last few years. In agreement with our results, Jaeger et al.[7] recently reported a mean asymptomatic interval of 3 years in 70 latex-allergic patients. Characteristic physical and chemical properties of the responsible allergens could, in part, explain this striking difference in the mean asymptomatic interval. According to several investigations, most of the different latex proteins are water soluble[7,10,12,20,21,40,65] and therefore may easily be released by sweat from the glove's surface, especially under occlusion. Furthermore, the high frequency of concomitant hand eczema in 60.6% of our 104 latex-allergic patients may promote the penetration of these water-soluble proteins. In contrast, the accelerators, which are known to cause most glove-related type IV allergies,[2,3,5,13,15,18,24] are almost water insoluble. Hence, they may only be released in a smaller amount, possibly leading to a prolonged asymptomatic interval. The lower rate of hand eczema (36.2% of 94 type IV-allergic patients) also may be a contributing factor.

In addition, the water solubility of latex proteins may, in part, be responsible for a lower risk of latex allergy in housekeepers compared to health care professionals (11.5 vs. 74% of our 104 latex-allergic patients). Because employees in medicine usually wear several pairs of gloves daily, they are exposed to a high amount of latex proteins, especially when using cheap examination gloves which often are poorly rinsed out during the manufacturing process. In contrast, the less well-fitting household gloves are normally used for a period of 4 to 6 weeks, which is presumably accompanied by a leaching of proteins and thus a continuous reduction of their allergenic potency.

According to our investigations, the increase of allergies to latex during the last few years seems to be accompanied by a shift to more severe forms of clinical manifestations (symptoms of a generalized allergy [stage II–IV] were found in 10.7 and 54.0% of all latex-allergic patients in 1989/1990 and 1991/1992, respectively). Therefore, the crucial clinical issue which should be addressed is the definition of main risk factors in order to reduce the frequency of latex allergies and to prevent systemic reactions in sensitized patients. Aside from occupations in health care service the presence of atopic diseases (mainly comprising allergic rhinitis) has been shown to predispose to latex allergy.[1,4,7,8,11,16,17,23,26,27,51,52,61] Hence, 63.5% of our 104 latex-allergic patients were atopics and 12.4% of 201 atopics suffered from a clinically relevant latex allergy. Lower prevalences of latex allergy (6.8%) among 44 atopic children were documented by Shield and Blaiss.[52]

Patients with spina bifida or congenital urologic anomalies, as found in two of our patients, are at special risk to develop latex allergy including anaphylactic shock reactions.[65] In a recent study, 69% of 26 children with spina bifida proved to have a positive latex RAST®[65] which presumably is due to frequent exposure to latex devices during surgery (gloves, catheters, enemas, etc.).

V. CONCLUSION

According to our investigations, a continuous overproportional increase in type I allergies to latex among health care professionals has to be taken into account in the future. In contrast, type IV allergies to rubber gloves presumably will be of minor importance, especially if thiurams are avoided as an ingredient. The identification of clinically relevant latex proteins and their elimination during the manufacturing process,[74,75] as well as the availability of reasonably inexpensive latex-free alternatives,[5] are necessary to stop this upward trend and to protect high risk patients. In addition, more knowledge about early symptoms of latex allergy among employees in medicine is mandatory for the prevention of systemic reactions.

REFERENCES

1. Beaudouin, E., Pupil, P., Jacson, F., Laxenaire, M. C., and Moneret-Vautrin, D. A., Allergie professionnelle au latex. Enquête prospective sur 907 sujets du milieu hospitalier, *Rev. Fr. Allergol.*, 30, 157, 1990.
2. Brandão, F. M., Rubber, in *Occupational Skin Disease*, Adams, R. M., Ed., W. B. Saunders, Philadelphia, 1990, chap. 28.
3. Estlander, T., Jolanki, R., and Kanerva, L., Dermatitis and urticaria from rubber and plastic gloves, *Contact Derm.*, 14, 20, 1986.
4. Fuchs, T. and Wahl, R., Allergische Soforttypreaktionen auf Naturlatex unter besonderer Berücksichtigung von Operationshandschuhen, *Med. Klin.*, 87, 355, 1992.
5. Heese, A., von Hintzenstern, J., Peters, K.-P., Koch, H. U., and Hornstein, O. P., Allergic and irritant reactions to rubber gloves in medical health services. Spectrum, diagnostic approach and therapy, *J. Am. Acad. Dermatol.*, 25, 831, 1991.
6. von Hintzenstern, J., Heese, A., Koch, H. U., Peters, K.-P., and Hornstein, O. P., Frequency, spectrum and occupational relevance of type IV allergies to rubber chemicals. A retrospective study from the department of Dermatology, University of Erlangen-Nuremberg, 1/1985–3/1990, *Contact Derm.*, 24, 244, 1991.
7. Jaeger, D., Kleinhans, D., Czuppon, A. B., and Baur, X., Latex-specific proteins causing immediate-type cutaneous, nasal, bronchial and systemic reactions, *J. Allergy Clin. Immunol.*, 89, 759, 1992.
8. Lagier, F., Vervloet, D., Lhermet, I., Poyen, D., and Charpin, D., Prevalence of latex allergy in operating room nurses, *J. Allergy Clin. Immunol.*, 90, 319, 1992.
9. Tomazic, V. J., Withrow, T. J., Fisher, B. R., and Dillard, S. F., Latex-associated allergies and anaphylactic reactions, *Clin. Immunol. Immunopathol.*, 64 (2), 89, 1992.
10. Alenius, H., Turjanmaa, K., Palosuo, T., Mäkinen-Kiljunen, S., and Reunala, T., Surgical latex glove allergy. Characterization of rubber protein allergens by immunoblotting, *Int. Arch. Allergy Appl. Immunol.*, 96, 376, 1991.
11. Axelsson, J. G. K., Johansson, S. G. O., and Wrangsjö, K., IgE-mediated anaphylactoid reactions to rubber, *Allergy*, 42, 46, 1987.

12. Carillo, T., Cuevas, M., Munoz, T., Hinojosa, M., and Moneo, I., Contact urticaria and rhinitis from latex surgical gloves, *Contact Derm.,* 15, 69, 1986.
13. Cronin, E., Rubber, in *Contact Dermatitis,* Cronin, E., Ed., Churchill Livingstone, Edinburgh, 1980, chap. 14.
14. Rycroft, R. J. G., Menné, T., Frosch, P. J., and Benezra, C., Eds., *Textbook of Contact Dermatitis,* 1st ed., Springer-Verlag, Berlin, 1992.
15. Frosch, P. J., Born, C. M., and Schütz, R., Kontaktallergien auf Gummi-, Operations- und Vinylhandschuhe, *Hautarzt,* 38, 210, 1987.
16. Frosch, P. J., Wahl, R., Bahmer, F. A., and Maasch, H. J., Contact urticaria to rubber gloves is IgE-mediated, *Contact Derm.,* 14, 241, 1986.
17. Kleinhans, D., Contact urticaria to rubber gloves, *Contact Derm.,* 10, 124, 1984.
18. Lammintausta, K. and Kalimo, K., Sensitivity to rubber. Study with rubber mixes and individual rubber chemicals, *Dermatosen,* 33, 204, 1985.
19. Lim, J. T. E., Goh, C. L., Ng, S. K., and Wong, W. K., Changing trends in the epidemiology of contact dermatitis in Singapore, *Contact Derm.,* 26, 321, 1992.
20. Mäkinen-Kiljunen, S., Turjanmaa, K., Palosuo, T., and Reunala, T., Characterization of latex antigens and allergens in surgical gloves and natural rubber by immunoelectrophoretic methods, *J. Allergy Clin. Immunol.,* 90, 230, 1992.
21. Morales, C., Basomba, A., Carreira, J., and Sastre, A., Anaphylaxis produced by rubber glove contact. Case reports and immunological identification of the antigens involved, *Clin. Exp. Allergy,* 19, 425, 1989.
22. Nutter, A. F., Contact urticaria to rubber, *Br. J. Dermatol.,* 101, 597, 1979.
23. Sussman, G. L., Tarlo, S., and Dolovich, J., The spectrum of IgE-mediated responses to latex, *J. Am. Med. Assoc.,* 265, 2844, 1991.
24. Taylor, J. S., Rubber, in *Contact Dermatitis,* Fisher, A. A., Ed., Lea & Febiger, Philadelphia, 1986, chap. 36.
25. Themido, R. and Brandão, F. M., Contact allergy to thiurams, *Contact Derm.,* 10, 251, 1984.
26. Turjanmaa, K., Incidence of immediate allergy to latex gloves in hospital personnel, *Contact Derm.,* 17, 270, 1987.
27. Turjanmaa, K., Laurila, K., Mäkinen-Kiljunen, S., and Reunala, T., Rubber contact urticaria. Allergenic properties of 19 brands of latex gloves, *Contact Derm.,* 19, 362, 1988.
28. Wrangsjö, K., Wahlberg, J. E., and Axelsson, I. G. K., IgE-mediated allergy to natural rubber in 30 patients with contact urticaria, *Contact Derm.,* 19, 264, 1988.
29. Moursiden, H. T. and Faber, O., Penetration of protective gloves by allergens and irritants, *Trans. St. John's Hosp. Dermatol. Soc.,* 59, 1, 1973.
30. Munksgaard, E. C., Permeability of protective gloves to (di)methacrylates in resinous dental materials, *Scand. J. Dent. Res.,* 100 (3), 189, 1992.
31. Storrs, F. J., Permanent wave contact dermatitis: contact allergy to glyceryl monothioglycolate, *J. Am. Acad. Dermatol.,* 11, 74, 1984.
32. Pegum, J. S., Penetration of protective gloves by epoxy resin, *Contact Derm.,* 5, 281, 1979.
33. Wall, L. M., Nickel penetration through rubber gloves, *Contact Derm.,* 6, 461, 1980.
34. von Krogh, G. and Maibach, H. I., The contact urticaria syndrome — 1982, *Semin. Dermatol.,* 1, 59, 1982.

35. Gronemeyer, W. and Debelic, M., Der sogenannte Reibtest, seine Anwendung und klinische Bedeutung, *Dermatologica (Basel)*, 134, 208, 1967.
36. Bonnekoh, B. and Merk, H. F., Safety of latex prick skin testing in allergic patients, *J. Am. Med. Assoc.*, 267 (19), 2603, 1992.
37. Wrangsjö, K., Mellström, G., and Axelsson, G., Discomfort from rubber gloves indicating contact urticaria, *Contact Derm.*, 15, 79, 1986.
38. De Corres, L. F., Munoz, D., Bernaola, G., Fernández, E., and Moneo, I. S., Sensitization to chestnuts and bananas in patients with contact urticaria from latex, *Contact Derm.*, 23, 277, 1990.
39. Lavaud, F., Cossart, C., Reiter, V., Bernard, J., Deltour, G., and Holmquist, I., Latex allergy in patient with allergy to fruit, *Lancet*, 339, 492, 1992.
40. Leynadier, F. and Dry, J., Allergy to latex, *Clin. Rev.*, 9, 371, 1991.
41. Leynadier, F., Pecquet, C., and Dry, J., Anaphylaxis to latex during surgery, *Anaesthesia*, 44, 547, 1989.
42. M'Raihi, L., Charpin, D., Pons, A., Bongrand, P., and Vervloet, D., Cross-reactivity between latex and banana, *J. Allergy Clin. Immunol.*, 87, 129, 1991.
43. Hjorth, N. and Roed-Petersen, J., Occupational protein contact dermatitis in food handlers, *Contact Derm.*, 2, 24, 1976.
44. Kleinhans, D., Soforttyp-Allergie gegen Latex: Kontakt-Urtikaria und Ekzem, *Akt. Dermatol.*, 10, 227, 1984.
45. Görtz, J. and Goos, M., Immediate and late type allergy to latex: contact urticaria, asthma and contact dermatitis, in *Current Topics in Contact Dermatitis*, Frosch, P. J., Dooms-Goossens, A., Lachapelle, J.-M., Rycroft, R. J. G., and Scheper, R. J., Eds., Springer-Verlag, Berlin, 1989, 457.
46. Rycroft, R. J. G., False reactions to nonstandard patch tests, *Semin. Dermatol.*, 5, 225, 1986.
47. Guillet, M. H., Ménard, N., and Guillet, G., Sensibilisation de contact aux gants en vinyl, *Ann. Dermatol. Venerol.*, 118, 723, 1991.
48. Wyss, M., Elsner, P., Wüthrich, B., and Burg, G., Allergic contact dermatitis from natural latex without contact urticaria, *Contact Derm.*, 28, 154, 1992.
49. Baur, X. and Jäger, D., Airborne antigens from latex gloves, *Lancet*, 335, 912, 1990.
50. Ring, J., *Angewandte Allergologie*, 2nd ed., MMV Medizin Verlag München, München, 1988.
51. Tarlo, S. M., Wong, L., Roos, J., and Booth, N., Occupational asthma caused by latex in a surgical glove manufacturing plant, *J. Allergy Clin. Immunol.*, 85, 626, 1990.
52. Shield, S. W. and Blaiss, M. S., Prevalence of latex sensitivity in children evaluated for inhalant allergy, *Allergy-Proc.*, 13 (3), 129, 1992.
53. Anderssen, K. E., Burrows, D., Cronin, E., Dooms-Goossens, A., Rycroft, R. J. G., and White, I. R., Recommended changes to standard series, *Contact Derm.*, 19, 389, 1988.
54. Van Ketel, W. G. and Van den Berg, W. H. H. W., The problem of the sensitization of dithiocarbamates in thiuram-allergic patients, *Dermatologica*, 169, 70, 1984.
55. Logan, R. A. and White, J. R., Carbamix is redundant in the patch test series, *Contact Derm.*, 18, 303, 1988.

56. Guimaraens, D., Gonzalez, M. A., and Condé-Salazar, L., Occupational allergic contact dermatitis and anaphylaxis from rubber latex, *Contact Derm.,* 26, 268, 1992.
57. Baptista, A., Barros, M. A., and Azenka, A., Allergic contact dermatitis on an amputation stump, *Contact Derm.,* 26, 140, 1992.
58. Cabrita, J. C., Goncalo, M., Azenka, A., and Goncalo, S., Allergic contact dermatitis of the eyelids from rubber chemicals, *Contact Derm.,* 24, 145, 1991.
59. Rademaker, M. and Forsyth, A., Carba mix: a useful indicator of rubber sensitivity, in *Current Topics in Contact Dermatitis,* Frosch, P. J., Dooms-Goossens, A., Lachapelle, J.-M., Rycroft, R. J. G., and Scheper, R. J., Eds., Springer-Verlag, New York, 1989, 136.
60. Turjanmaa, K., Reunala, T., and Räsänen, L., Comparison of diagnostic methods in latex surgical glove contact urticaria, *Contact Derm.,* 19, 241, 1988.
61. Baur, X., Jäger, D., Engelke, T., Rennert, S., and Czuppon, A. B., Latexproteine als Auslöser respiratorischer und systemischer Allergien, *Dtsch. Med. Wochenschr.,* 117, 1269, 1992.
62. Ehl, W., Hartjen, A., Thiel, C. L., Aulepp, H., and Fuchs, E., Latex-Allergien als IgE-vermittelte Sofortreaktionen, *Allergologie,* 11 (5), 182, 1988.
63. Foussereau, J., Brändle, I., and Boujnah-Khouadja, A., Allergisches Kontaktekzem durch Isothiazolin-3-on-Derivate, *Dermatosen,* 32, 208, 1984.
64. Milcovic-Kraus, S., Glove powder as a contact allergen, *Contact Derm.,* 26 (3), 198, 1992.
65. Slater, J. E. and Chhabra, S. K., Latex antigens, *J. Allergy Clin. Immunol.,* 89 (3), 673, 1992.
66. Assalve, D., Cicioni, C., Perno, P., and Lisi, P., Contact urticaria and anaphylactoid reaction from cornstarch surgical glove powder, *Contact Derm.,* 19, 61, 1988.
67. Fisher, A. A., Contact urticaria and anaphylactoid reaction due to corn starch surgical glove powder, *Contact Derm.,* 16, 224, 1987.
68. Seggev, J. S., Mawhinney, T. P., Yunginger, J. W., and Braun, S. R., Anaphylaxis due to cornstarch surgical glove powder, *Ann. Allergy,* 65, 152, 1990.
69. Belsito, D. V., Contact urticaria caused by rubber: analysis of seven cases, *Dermatol. Clin.,* 8, 61, 1990.
70. Camarasa, J. M. G. and Alomar, A., Allergic rhinitis from diphenyl guanidine, *Contact Derm.,* 4, 242, 1978.
71. Helander, I. and Mäkelä, A., Contact urticaria to zinc diethyldithiocarbamate (ZDC), *Contact Derm.,* 9, 327, 1983.
72. Lahti, A. and Maibach, H. I., Immediate contact reactions, in *Occupational and Industrial Dermatology,* 2nd ed., Maibach, H. I., Ed., Year Book Medical Publishers, Chicago, 1987, chap. 6.
73. Viana, I. and Brandão, F. M., As luvas como factor de agrovamento de algumas dermatoses profissionalis, *Trab. Soc. Port. Dermatol. Venerol.,* 46, 85, 1988
74. Dalrymple, S. J. and Audley, B. G., Allergenic proteins in dipped products: factors influencing extractable protein levels, *Rubber Dev.,* 45, 51, 1992.
75. Leynadier, F., Xuan, T. T., and Dry, J., Allergenicity suppression in natural latex surgical gloves, *Allergy,* 46, 619, 1991.

14

Protective Effect of Gloves Illustrated by Patch Testing — Practical Aspects

Carola Lidén and Karin Wrangsjö

TABLE OF CONTENTS

I. Introduction 207
II. Patch Testing With Allergens and Gloves 208
III. Practical Aspects 209
IV. Conclusions 211
References 212

I. INTRODUCTION

One main task in occupational dermatology is to give correct information and advice concerning protective gloves. The skin protection afforded by gloves depends on several factors. The primary factor is the effectiveness of the barrier function of the glove material itself, including the ways the material is affected by chemicals during use. Factors that may influence the effectiveness of skin protection arise when gloves are reused over periods of time. The chemical contamination of both surfaces constitutes a real risk factor. Permeation through protective gloves by many chemicals such as acrylate monomers, epoxy resins, and glycerol monothioglycolate is well documented and described in other chapters of this book. Glove storage routines, including storage time and temperature, cleaning, and time in use may all influence the protection the gloves give to the skin.

For clinical situations where there may be limited accurate data on chemical permeation through special types and brands of protective gloves, patch testing with actual allergens applied on pieces of the gloves may be a useful part of the examination.

Likewise, in some clinical situations in occupational dermatology, correct patch testing includes different separate tests with the suspected allergens and comparison of the results of patch testing with new and used gloves. Some of these practical aspects of the testing routines will be discussed in this chapter.

II. PATCH TESTING WITH ALLERGENS AND GLOVES

The protective effect of gloves may be illustrated by patch-testing of contact-allergic individuals with the specific allergen together with pieces of glove. This has been done in a few studies, e.g., with photographic chemicals, epoxy, and glyceryl monothioglycolate.[1-3] A positive patch-test reaction produced through a glove material indicates that the allergen has passed through the material, either by penetration or by permeation.[4] The patch test method may be used when no data on penetration or permeation are available, or when chemical analysis for detection or quantification of the substance is not possible.

In a limited study, five individuals with known contact allergy to some photographic chemicals were patch tested with one or two allergens in combination with pieces of five different protective gloves. The study was part of a field study concerning occupational skin disease in a film laboratory.[1] The workers, especially the mixers, developers, and analysts, were heavily exposed to chemicals, many of which are known contact allergens or skin irritants. The prevalence of occupational skin disease, mainly contact allergy, was high.

The gloves were of the types used at the film laboratory where the study was carried out, and of the types used at other film laboratories visited (Table 1). The allergens were the color-developing agents CD-2 and CD-3, the black-and-white developer Metol, and a bleach accelerator, PBA-1, which had recently been described as a contact allergen. All were tested at 1% in water. Patch testing was carried out in the following way: between the skin of the back and the test patch (Al-test, Imeco, Sweden) with each allergen, a 35 × 35 mm piece of each glove was placed. As a positive control, every substance was tested without the different glove materials. Each person was also tested with the five glove materials, with a test patch but without allergen, as negative controls (Figures 1 and 2).

The results are shown in Table 2. CD-2, CD-3, and Metol caused patch-test reactions through the disposable gloves made of latex, PVC, and polyethy-

TABLE 1   Gloves Used in the Study

Material	Type	Name and producer
Latex	Household	Fagra, Imported from France
Latex	Disposable	White Knight Examtex, Imported, unknown
PVC	Household	Pentry, AB Vinylprodukter, Sweden
PVC	Disposable	Tru-Touch, Edmont, Belgium
Polyethylene	Thin, disposable	LIC-Disposable, Imported from Hong Kong

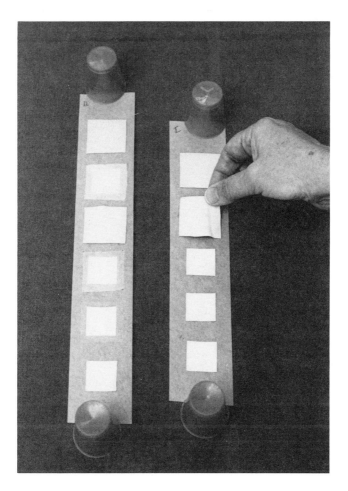

FIGURE 1 Preparing patch test with allergen (CD-2) and pieces of gloves: Scanpor, Al-test, allergen and pieces of gloves (35 × 35 mm).

lene. CD-2 and Metol also caused reactions through the thick latex and PVC gloves. PBA-1 did not cause any patch-test reaction through the different glove materials tested. The study being very limited, definite conclusions concerning the protective effect of the gloves tested cannot be drawn. That most of the chemicals studied did pass through most of the gloves tested, however, was illustrated in a fairly simple way. Other methods are needed to investigate mechanism (penetration or permeation), breakthrough time, etc.

III. PRACTICAL ASPECTS

Many patients examined for suspected occupational dermatitis are patch-tested with various materials from their work. To be complete, this patch-test inves-

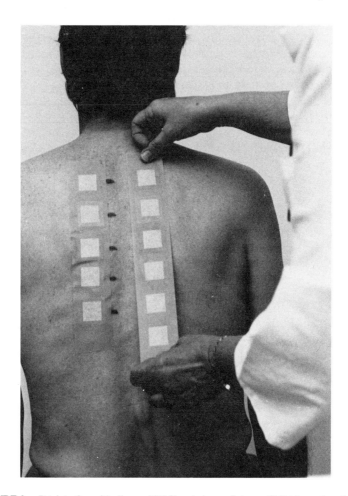

FIGURE 2 Patch testing with allergen (CD-2) and pieces of gloves. (I) Patch numbers 1 through 5 (left side): Al-test with no allergen and with pieces of five different gloves (negative controls). (II) Patch numbers 1 through 5 (right side): allergen on Al-test and pieces of five different gloves; number 6: allergen on Al-test (positive control).

TABLE 2 Results from Patch Testing With Allergens (1% In Water) and Pieces of Protective Gloves in Five Individuals With Contact Allergy to Photographic Chemicals

Allergen (no. tested)	Glove type					
	Latex household	Latex disposable	PVC household	PVC disposable	Polyethylene	No glove
CD-2 (n = 3)	1	2	1	1	3	3
CD-3 (n = 2)	0	1	0	1	1	2
Metol (n = 2)	1	2	1	2	2	2
PBA-1 (n = 2)	0	0	0	0	0	2
No allergen (n = 5)	0	0	0	0	0	—

tigation includes tests with the different types of glove used by the patient. The patients are asked to bring the different gloves, if possible still packed. There is as yet no requirement to label the latex or chemical content in gloves, but in most cases the type of material is stated.

Because different chemicals may contaminate, penetrate, and permeate glove materials, there are reasons to compare the results of testing used and unused gloves. For the same reason, we also recommend testing both surfaces of the gloves. This latter recommendation is justified by the fact that in the development of chemical-resistant gloves and new types of "hypoallergenic" gloves, different materials are used for the outer surfaces and the inner surfaces.

For patch testing we apply 3×3 cm pieces cut from the glove materials on the patient's back. Application time is 48 h, with reading at 72 h, according to international recommendations.[5]

We have found good correlation between the patch-test results and the clinical picture in cases of rubber allergy. When PVC and polyethylene materials are tested, most patients react with weak erythema on the test areas. The test results are the same in control patients with PVC and polyethylene glove materials. This indicates that these materials in most cases provoke reactions which are probably irritative rather than of true allergic origin.

When patch-testing the chemicals and the gloves from the patients' actual workplaces in parallel, one sometimes finds the offending allergens even inside the protective gloves used. This is usually of high educational value and stimulates discussion of the choice, use time, and storage of gloves.

The presence of allergens in gloves may also be demonstrated by chemical analyses. The dimethylglyoxime test for nickel is useful, not only for demonstrating nickel on metal surfaces, but on glove surfaces as well.[6] With this test, we have found nickel contamination on rubber gloves used in the metal-plating industry. Strong positive results with the dimethylglyoxime test also indicated heavy nickel contamination of the inner surface of the gloves.

The presence of formaldehyde in glove materials may be demonstrated with the chromotropic acid test.[7] With this method we have found formaldehyde in thin cotton gloves often used by our patients with hand dermatitis, or used under household gloves at work. The chromotropic acid test is qualitative, and positive results from cotton gloves should be supplemented with more accurate qualitative tests. Still, because many textiles are treated with formaldehyde resins, and because our test results point to the possibility of release of formaldehyde from cotton glove materials, we recommend that patients with hand dermatitis wash their cotton gloves before the first use.

IV. CONCLUSIONS

To handle clinical problems which may arise from penetration, permeation, or contamination of gloves by chemicals used at work, the following patch test routines may be recommended:

1. Patch test with pieces from used and unused gloves
2. Patch test with the outer and the inner surface
3. Consider patch testing with the suspected allergen and a piece of glove material together
4. Simple chemical analyses may be carried out to demonstrate some contaminants on the gloves, such as nickel and formaldehyde

REFERENCES

1. Lidén, C., Occupational dermatoses at a film laboratory, *Contact Derm.,* 10, 77, 1984.
2. Jolanki, R., Estlander, T., and Kanerva, L., Contact allergy to an epoxy reactive diluent: 1,4-butanediol diglycidyl ether, *Contact Derm.,* 16, 87, 1987.
3. McClain, D. C. and Storrs, F. J., Protective effect of both a barrier cream and a polyethylene laminate glove against epoxy resin, glyceryl monothioglycolate, Frullania, and Tansy, *Am. J. Contact Derm.,* 3, 201, 1992.
4. Mellström, G., Protective Gloves of Polymeric Material. Experimental Permeation Testing and Clinical Study of Side Effects, Karolinska Institute, Ph.D. thesis, *Arbete och Hälsa,* 10, 1991.
5. Wahlberg, J. E., Patch testing, in *Textbook of Contact Dermatitis,* Rycroft, R. J. G., Menné, T., Frosch, P. J., and Benezra, C., Eds., Springer-Verlag, Berlin, 1992, chap. 10.
6. Fisher, A. A., *Contact Dermatitis,* 3rd ed., Lea & Febiger, Philadelphia, 1986, 752.
7. Fisher, A. A., *Contact Dermatitis,* 3rd ed., Lea & Febiger, Philadelphia, 1986, 296.

PART V

SIDE EFFECTS WHEN USING PROTECTIVE GLOVES

15

Irritation and Contact Dermatitis from Protective Gloves — An Overview

Jan E. Wahlberg

TABLE OF CONTENTS

I. Introduction 215
II. History 216
III. Symptoms and Signs 216
IV. Exposure Conditions 216
V. Skin Irritation — Strategy for Examination and Suggested
   Investigations 217
   A. Provocation Tests 217
   B. Patch Tests With Pieces of Glove 218
   C. Workplace Visits 218
VI. Summary and Conclusion 218
References 219

I. INTRODUCTION

The beneficial value of using protective gloves varies, but for most people the advantages outweigh the disadvantages. Side effects on the skin of users' hands and forearms unfortunately are not uncommon. In a Japanese questionnaire investigation[1] among 792 housewives, 456 used gloves and 61 (13%) reported skin troubles. In some cases the adverse effects are so severe that users have to abandon these excellent prophylactic devices.

In any single case it is hard to predict the interaction and relative importance of exposure to chemicals and products, a preexisting hand dermatosis with or without impaired barrier function, the glove materials, and the occlusive effect of the gloves.

Some common and more or less well-defined side effects are presented in Table 1. Allergic contact dermatitis from glove materials, contact urticaria and glove dermatitis of other types are reviewed in detail in this book in Chapters

TABLE 1 Side Effects on the Skin When Using Protective Gloves — An Overview

Effect	Discussed in
Allergic contact dermatitis from ingredients/components in gloves made of rubber or plastic	Chapter 16
Contact urticaria from latex gloves	Chapter 17
Glove dermatitis — other reactions: irritation (glove powders, occlusion, maceration, etc.), other urticaria, chemical leucoderma, endotoxins, ethylene oxide	Chapter 18

16, 17, and 18. However, there still remain patients in whom no specific skin disease can be diagnosed despite the extensive and detailed examination procedures described in these and Chapters 10 through 14.

In this chapter, some general aspects are presented of the examination and management of those cases where a specific cause or a defined skin disease — according to Table 1 — have been excluded.

II. HISTORY

A careful history is essential for disclosing atopy/atopic hand eczema, hyperhidrosis/pompholyx, psoriasis, tinea, dermographism, or any other hand dermatoses that might be worsened by the occlusive effect of the gloves. Were gloves first used due to a preexisting skin disease, or a general recommendation from a supervisor, or on the patient's own initiative?

III. SYMPTOMS AND SIGNS

The patient's usual report of itching, burning, redness, edema, fissuring, and scaling is of minor value when deciding whether the complaints are glove-related. Location on the backs of the hands and fingers, with a sharp cut-off dermatitis at the wrists, suggests that the condition is glove-related.[2] Glove materials will rarely cause side effects on the volar aspects of the hands, solely. The distribution, then, will be of some help with the diagnosis, while the morphology of the lesions may mimic several other hand dermatoses.

Some patients referred to us were diagnosed as having scabies, pustulosis palmaris, or hand psoriasis, and had incorrectly blamed the gloves they had used. Differential diagnoses are discussed in Chapters 16, 17, and 18.

IV. EXPOSURE CONDITIONS

The relative contribution of the following factors must be considered:

> The glove materials (with and without linings): rubber, plastic, leather, textile, wire cloth, metal mesh, etc., or mixtures of these. Each type has its own area of use and pattern of skin side effects.

What chemicals and products were handled? Are any of these known to be irritants, allergens, and/or permeable through glove materials? (See Chapters 6 through 9.) Once the penetrants are inside the gloves, their harmful effects are enhanced by occlusion, humidity, and skin temperature. Some solvents and tensides can also act as extractants and vehicles and bring ingredients of glove materials to the skin surface, thereby facilitating the induction of irritation and allergy.

Were the gloves accidentally contaminated through their openings? Such contamination might give rise to high concentrations of a chemical or product inside the glove and — if undetected — cause skin irritation augmented by occlusion.

Were the gloves used during the entire exposure period?

Were there periods of exposure with unprotected hands — at work, at home, or in leisure time? Patients might wrongly blame the gloves they have used.

How often were the gloves changed? Were new pairs regularly, or were the contaminated gloves cleaned and reused? After removal of the gloves, the diffusion of chemicals through the glove membrane will continue and the chemicals will be trapped — unless the gloves are cleaned. On the following day the threshold of irritancy of a particular agent might have been reached and the subject will experience subjective and objective signs of skin irritation.

Has a recent change taken place in brands of gloves? (Batch-to-batch variation.)

Were any other people affected simultaneously after using gloves from the same supplier or delivery?

V. SKIN IRRITATION — STRATEGY FOR EXAMINATION AND SUGGESTED INVESTIGATIONS

The tests necessary for confirming the diagnoses of contact allergy and contact urticaria are reviewed in Chapters 16 and 17. It is essential to detect these cases early in order to initiate the appropriate prophylactic approach and then in follow-up consultations to check the course and outcome.

Some of the side effects described in Chapter 18 cannot be verified by any objective tests or analyses — the diagnoses are then based essentially on exclusion criteria.

The situation for skin irritation is similar — there is no confirmative test or analysis and the diagnosis cannot be settled until other alternatives have been excluded. Some common causes of skin irritation in glove users are summarized in Table 2.

A. Provocation Tests

Patients are asked to bring all suspected gloves to the clinic and provocation tests with the whole intact glove on moistened skin are carried out under supervision. In some cases, 1 to 3 h of exposure is insufficient for eliciting symptoms or rashes; the test must be repeated on subsequent days.

TABLE 2 Some Common Causes of Skin Irritation in Glove Users

Occlusion, sweating, and maceration
Enhanced penetration of irritants from, e.g., detergents and skin care products due to the occlusion
Agents that have penetrated the glove membranes from the outside (environmental exposure)
Contamination of the inside by accidental entrance through the glove opening
Agents/materials emanating from the glove itself
Agents/materials remaining from the glove manufacturing process
Agents remaining from the glove washing procedure — detergents and moisture
Glove powders, ethylene oxide, etc. reviewed in Chapter 18
Friction from the seams of leather and textile gloves[3]

B. Patch Tests With Pieces of Glove

Patch tests with pieces of gloves have probably already been carried out when looking for contact allergy (see Chapter 11) and for contact urticaria (see Chapter 17). When a patient reacts only to such pieces, but not to the rubber allergens in the standard series, this may indicate glove-related skin irritation. However, hundreds of rubber chemicals are known, and the standard patch test series can only contain the most important allergens among these.

Simultaneous patch tests are recommended, with new (unused) and used (i.e., contaminated) glove pieces (with the inside as well as the outside glove surface toward the skin). The results should then be compared. A reaction only to the used glove membrane indicates that the reactivity might be related to other factors than the glove itself.

C. Workplace Visits

If the tests discussed above fail to reproduce the symptoms and signs of irritation, a workplace visit is recommended. In some cases the riddle is solved during such a visit, and the gloves were found not to be the culprit.

VI. SUMMARY AND CONCLUSION

Skin irritation cannot be diagnosed by any specific test: a "diagnosis of exclusion" (Table 3) is required. It is essential to diagnose contact allergy, contact urticaria, and other types of glove dermatitis (see Chapters 16, 17, and 18) since these are preventable.

To support the diagnosis of skin irritation, we recommend provocation tests with whole gloves, patch tests with pieces of the glove, and workplace visits, especially in those cases where contact allergy, contact urticaria, etc., have been ruled out.

TABLE 3   Skin Irritation From Glove Use — A Diagnosis of Exclusion

Exclude contact allergy and contact urticaria (according to the criteria presented in Chapters 16 and 17)
Exclude other types of glove dermatitis (presented in Chapter 18)
Supervised provocation tests at the clinic, using suspected gloves
Patch tests with used and unused glove pieces; both inside and outside of glove

REFERENCES

1. Aoyama, M., Sugiura, K., Fujise, H., and Naruse, M., On use of gloves in the home and their influence upon skin irritation, *Nagoya Med.*, 27, 65, 1982.
2. Fisher, A. A., Management of dermatitis due to surgical gloves, *J. Dermatol. Surg. Oncol.*, 11, 628, 1985.
3. Estlander, T. and Jolanki, R., How to protect the hands, *Dermatol. Clin.*, 6, 105, 1988.

16

Allergic Contact Dermatitis from Rubber and Plastic Gloves

Tuula Estlander, Riitta Jolanki, and Lasse Kanerva

TABLE OF CONTENTS

I. Introduction 221
II. Frequency of Allergic Contact Dermatitis Due to Gloves 222
    A. Rubber Gloves 222
    B. Plastic Gloves 224
III. Sensitizers 224
    A. Rubber Gloves 224
    B. Plastic Gloves 228
IV. Clinical Aspects 229
    A. Predisposing Factors 229
    B. Sites and Appearance of Dermatitis 230
V. Diagnosis and Differential Diagnosis of Glove Dermatitis 232
VI. Prevention of Glove Dermatitis 233
VII. Summary 234
References 234

I. INTRODUCTION

Protective gloves made of polymeric rubber and plastic materials are important, although only secondary, safeguards against factors hazardous to hands. Gloves give the best possible protection in situations where the hands cannot be protected by more effective technical measures. They are used to protect the hands mainly from chemical and biologic agents, but they can also be used against physical and mechanical hazards. Polymer gloves may also be necessary for the protection of products against workers' dirty hands or in the protection of patients against contamination by microbes on the hands of medical personnel.

Although the use of polymer gloves often may be the only way to protect the hands against occupational and nonoccupational environmental hazards, or even modify their effects, their use also entails other problems.[1] The development of allergies to glove materials and possibly other agents used in gloves is probably the most harmful disadvantage induced by polymer gloves. Allergy, once developed, is generally a permanent state of the organism. When a sensitized person comes into contact with the specific sensitizer(s), clinical symptoms of allergy, e.g., dermatoses, usually recur.

Rubber gloves may induce both delayed allergy or type IV allergy leading to contact eczema (dermatitis), and immediate allergy or type I allergy appearing as contact urticaria or as immediate or protein contact dermatitis (see Chapter 17). Allergic contact dermatitis (type IV allergy) caused by rubber gloves is possibly more common than type I allergic contact urticaria.[2-4] In certain occupations, however, where natural rubber gloves are used daily for long periods, contact dermatoses due to type I allergy may be more important.[5-7] Simultaneous type I and IV allergy to rubber gloves may also occur.[5,6,8]

Plastic gloves usually cause allergic contact dermatitis as a result of the development of type IV allergy in response to the glove materials.[9,10] Type I reactions to plastic materials are rare.[11,12] So far, no cases have been reported from plastic gloves.

II. FREQUENCY OF ALLERGIC CONTACT DERMATITIS DUE TO GLOVES

Most cases of allergic contact dermatitis (type IV allergy) induced by polymer gloves are connected with the use of rubber gloves.[2-4,10,13-28] Most of these cases are induced by gloves made of natural rubber latex (NRL), whereas cases reported from synthetic rubber gloves are rather few. Plastic gloves have been the cause of allergic contact dermatitis only in a few reported cases. Polyvinylchloride (PVC, vinyl) plastics have been responsible for the sensitization in these cases.[9,10,29-35]

A. Rubber Gloves

The use of rubber gloves began early in the 1900s.[26] As easily available protective devices, they have become familiar to everyone needing hand protection at work, at home, or in pursuing hobbies. In the decade or two after their introduction, it became apparent that rubber products can induce contact dermatitis. Rubber gloves, especially surgical gloves (reviewed by Bonnevie and Marcussen in 1944[13] and Schultheiss et al. in 1959[15]), and also industrial rubber gloves[36] were among the first rubber products reported to have evoked contact allergy to rubber. Since the end of the 1950s with the increased manufacture and use of rubber articles for different purposes, including protective devices, gloves became even more a common cause of allergic contact rubber eczema.[14-17,37] Nowadays, most cases of allergic rubber eczema are caused by rubber articles[2,23,24,27] and gloves are considered to be the main source of occupational as well as nonoccupational sensitization to rubber.[3,4,9,18-21,24,38,39]

In a Spanish study covering the period 1973 to 1977,[39] of 2784 contact dermatitis patients, 338 (12%) were allergic to rubber ingredients. Of 262 occupationally related cases (housewives included), 95% were sensitized from the use of rubber gloves, and about half of these were due to household gloves worn by housewives. Of the rest of the cases, 30% were due to gloves in industrial use, 5% from the gloves of health care personnel, and 10% from hairdressers' gloves or gloves worn in other occupations.

Similar results were reported from a Portuguese clinic: 313 (7%) out of 4564 patients from 1977 to 1982 were allergic to rubber ingredients. An occupational source (domestic work included) was considered as being responsible for the sensitization in 191 patients (62%), and almost all of them (96%) were sensitized from the use of gloves. Masons were the largest occupational group sensitized in their work (117 = 64%).[21] In an Australian study, 50 (10%) of 486 patients investigated in a contact dermatitis clinic from 1976 to 1978 were allergic to rubber, but rubber gloves had been the agent in 68% of the cases.[18] Finnish studies on patients investigated between 1974 and 1983 in an industrial dermatology clinic have given similar results. Out of 1052 patients who had occupational contact dermatitis, 108 (10%) were allergic to rubber.[40] Rubber gloves were the source of the sensitization in 58% of the cases.[9] When the period from 1974 to 1991 is considered, the proportion of rubber gloves as the cause of occupational allergic rubber eczema was the same (Table 1). In 1983 to 1988, 25% of the cases of allergic contact eczema, reported from the entire country to the Finnish Occupational Disease Register, were due to rubber.[41] During the 1990 to 1991 period, the proportion was 28%.[42] Rubber gloves are considered to be the main source of occupational sensitization to rubber throughout the country; they are estimated to have induced two thirds of the reported cases.

A somewhat lower proportion for rubber gloves, 41%, was found in a Belgian study where 55 (7%) out of 810 patients tested for 1 year were sensitized to rubber.[19] In a recent German study of 3851 patients diagnosed as

TABLE 1 Allergic Dermatoses Due to Rubber (n = 190) 1974–1991, Institute of Occupational Health, Helsinki

Cause	Allergic eczema	Contact urticaria	%
Gloves	98	13	58
Black rubber	45	0	24
Other rubber	34	0	18
Total	177	13	100

having allergic contact eczema between 1985 and 1990, only 145 (4%) were sensitized to rubber ingredients — 80 (55%) of these 145 cases were of occupational origin and 84% were caused by NRL gloves.

Variations in the results reported from different countries probably reflect differences in the examined patients (i.e., selectivity of patients in contact dermatitis and occupational dermatology clinics), glove materials used, the ways in which the gloves are used, numbers of patients investigated, as well as differences in test techniques and substances used to detect the sensitization.

B. Plastic Gloves

Completely cured plastic materials are not generally considered to be sensitizers. PVC, polyethylene (PE), polyvinylacetate (PVAC), and polyvinylalcohol (PVA) and other materials used in plastic gloves rarely cause allergic contact dermatitis.[43] The use of plastic materials for personal protective equipment did not become common until in the 1950s.[44] Plastic gloves, even nowadays, are possibly less frequently used than rubber gloves. One reason for this is that the users often find plastic gloves less comfortable because they are not always as soft and pliable as rubber gloves and do not completely follow the contours of the hands. Accordingly, most of the reports of allergic contact eczema from plastic gloves are based on only one to five patients.

In a German study, 31 patients investigated during 1969 to 1984 were sensitized from the use of rubber or vinyl gloves; 10% of them were allergic to vinyl gloves.[10] Similar results were also obtained in a Finnish study: 5 (7%) of 68 patients were sensitized from the use of PVC gloves.[9] Since then, no new definitive cases of plastic glove allergy have been diagnosed at our clinic. Some suspicions have arisen but, to date, irritation has remained as the cause of glove discomfort. In a Japanese study,[45] however, 31 (51%) out of 61 women who had developed contact dermatitis from using household gloves connected their skin symptoms with the use of vinyl gloves, and 26 (43%) with rubber gloves. In four cases the type of glove used was not clear. Irritation was suggested as the cause of the vinyl glove dermatitis, not allergy to the material itself. The discomfort caused by vinyl gloves is not uncommon in Japan. Vinyl gloves are frequently worn by housewives, possibly because Japanese investigators have suggested that vinyl gloves provide better protection against contact dermatitis.[46] Recently, a fraction, probably an irritant in the vinyl gloves sold in Japan, has been discovered in animal tests.[47]

III. SENSITIZERS

A. Rubber Gloves

Rubber and plastic gloves are usually manufactured by various automated processes, of which the dipping method is the most common.[48] Rubber gloves

TABLE 2 Household Glove Dip Made of Natural Rubber Latex

Component	Dry Parts
Centrifuged NRL (60%)	100
Nonionic stabilizer solution (20%)	0.5
KOH solution (10%)	0.25
Sulfur dispersion (50%)	1.75
Zinc mercaptobenzothiazol dispersion (50%)	1
Antioxidant dispersion (40%)	0.5
Zinc oxide dispersion (40%)	2

From Morton, M., *Rubber Technology,* Van Nostrand Reinhold, New York, 1973. With permission.

are prepared from NRL or synthetic rubber polymers to which many chemicals are added during the manufacturing process. The mixture may contain additives, such as vulcanizing agents, accelerators, antioxidants, dyes, pigments, fillers, and oils.[49,50] A list of ingredients used in the manufacture of a household glove made of NRL[49] is given in Table 2. Examples of synthetic materials are Neoprene® (polychloroprene) and nitrile rubbers. The gloves may also have been prepared of mixed rubber materials, e.g., nitrile or Neoprene® rubber and NRL polymers. The final rubbery properties of the gloves are obtained during the vulcanization process. During the process some of the chemicals added to the mixture may degrade partially or totally, and new compounds may develop in the material.[23,50] Glove powders are other important agents used, especially in surgical and examination gloves (for details see Chapter 3).

The additives, not the rubber polymers themselves, are the chief sensitizers in rubber gloves (Table 3). Reports on cases induced by rubber polymers are rare. Apart from a case caused by isoprene in a rubber factory,[23] there are only a few previous descriptions pointing to allergic contact dermatitis from NRL polymers in gloves: a case of Bonnevie and Marcussen from 1944,[13] and two other cases reviewed by Cronin.[20] Recently, three new cases[34,51,52] have been reported suggesting that NRL polymers are the cause of contact allergic eczema from NRL gloves. Nevertheless, stabilizers and preservatives in NRL liquids and new compounds possibly formed in the material during the vulcanization process,[3,4,50] as well as additives not included in common rubber test series, have not totally been excluded as causes of the glove reactions in these cases.

Small quantities of certain additives, such as accelerators and antioxidants, may not become structural components of the rubber substance[23,53] and may therefore be present in a relatively free state in the material. The majority of the cases of rubber glove allergy may be connected with the accelerators. Thiurams, benzothiazoles, carbamates, and thioureas are the responsible chemicals; most of the allergy cases have been caused by the first three groups of these

TABLE 3 Sensitizers in Rubber Gloves

Main
  Accelerators
    Thiurams
    Carbamates
    Benzothiazoles
    Thioureas
Other
  Antioxidants
    p-Phenylenediamine derivatives
    Phenol derivatives
  Vulcanizers
  Colorants
    Organic pigments
  Preservatives
  Antistatic agents
Potential
  Donning powders
  Fragrances

chemicals.[3,4,9,10,13,15-22,26,27,54-56] Mercaptobenzothiazole (MBT), the first benzothiazole accelerator introduced into the rubber industry during the 1920s, was also the first chemical considered as being the main type IV allergen in rubber gloves,[13,54] although reactions to thiurams also occurred.[16,17,54] Some reports connect allergy to MBT, especially in surgical gloves,[27,55] while others[57] consider thiurams as the chief allergens among operating room staff. Nowadays, MBT can be regarded as the third most common allergen, after thiurams and carbamates, in glove allergic patients.[2-4,20,27] Other allergenic derivatives of MBT can also be used as accelerators, e.g., dibenzothiazyl disulfide (MBTS), morpholinyl mercaptobenzothiazole (MMBT), and N-cyclohexylbenzothiazyl sulfenamide (CBS).[58] An analysis of 11 NRL gloves revealed CBS in some of the gloves.[59] In addition, commercial grade of MBT can be contaminated with CBS.[60] Other disulfide-containing derivatives of MBT can be even more sensitizing than MBT.[60-62] It has also been shown experimentally that, depending on environmental factors, MBT, MMBT, and CBS can be converted to MBTS (oxidizing milieu) and back again to MBT (reducing milieu).[58]

Thiurams and carbamates are frequently used in gloves.[3,4,10,16,20,28,56,63] Correspondingly, thiurams have most commonly been responsible for the reactions due to rubber gloves — the frequencies ranging from 54 to 80%.[3,4,9,17-22,56] Dipentamethylenethiuram disulfide,[16,17,20] tetraethylthiuram disulfide, and tetramethylthiuram monosulfide and disulfide have been the single most frequent reactors among the thiurams, possibly reflecting variations in the glove materials in different countries.[3,4,9,10,21,27,56]

Carbamates are the second most common cause of allergic reactions in glove-allergic patients.[3,4,9] Simultaneous reactions to thiurams and related carbamates are usual, owing to concomitant sensitization or cross-reactions between the chemicals.[20,23,56,63] One of the most frequent carbamate reactors is zinc diethyldithiocarbamate (ZDC),[3,4,9,10,17,22,56] but other carbamates, e.g., zinc dimethyl, dibutyldithiocarbamates, and piperidine pentamethylenedithiocarbamate can also be used in gloves.[16,48,59] A case of sensitization to ZDC was reported in 1981.[64] Recently, another case pointing to pentamethylenedithiocarbamate or piperidine as the cause of glove dermatitis has also been published. These chemicals were found in three different brands of surgical gloves.[65] Carbamates and benzothiazoles, possibly considered as less sensitizing accelerators than thiurams, are often used in hypoallergenic surgeons' and examination gloves made of NRL or synthetic rubber polymers.[3,4,63,66]

Thiourea derivatives are rather new allergenic accelerators found in Neoprene® rubber gloves. Diphenylthiourea is one of the thioureas used in these type of gloves.[67,68] Classified as moderate sensitizers,[67] they can also be found in hypoallergenic surgical gloves.[3,4] Reports on sensitization to thioureas in gloves are few. A case of allergy to dihydroxydiphenyl (DOD) and diphenylthiourea in Neoprene® gloves has been reported.[68] In addition, another report describes two patients suspected of having rubber glove dermatitis. Positive reactions were found to diethyl-, diphenyl-, and dibutylthioureas.[69] Also, one patient at our clinic, probably having been sensitized from the use of a certain brand of industrial Neoprene® rubber gloves and rubber boots, was patch test positive to diphenylthiourea.[98] Recently, we reported on two other patients who probably had been sensitized to the diphenylthiourea in their Neoprene® protective gloves.[99] DOD is a rare sensitizer,[20] but it has been the cause of glove dermatitis in four surgeons and one nurse.[70] Two of the glove-allergic patients at our clinic were also positive to DOD.[9]

Antioxidants are less frequent sensitizers in gloves. *p*-Phenylenediamine derivatives, which may be strong sensitizers, are seldom used in rubber gloves except in black or dark colored industrial gloves.[9,25,59,71] Some cases of sensitization to *N*-isopropyl-*N*-phenyl-*p*-phenylenediamine (IPPD) in gloves have been published.[9,25,72,73] Other antioxidants which have only rarely caused sensitization in gloves include phenol derivatives. These are used in all kinds of rubber gloves.[16,28,59,71] An analysis of antioxidants in 37 different brands of commercially available rubber gloves in Japan for domestic use revealed four phenolic antioxidants, of which 2,2'-methylene-*bis*(4-methyl-6-*tert*-butylphenol) (MBMBP) was most frequently present. Patch tests to 40 rubber-sensitive patients were negative with MBMBP and with another antioxidant.[71] Wilson,[16] however, found one person who was allergic to MBMBP in his series of rubber glove-sensitive patients. Another phenol derivative, 2-*p*-methylcyclohexyl-*p*-4,6-dimethylphenol, has also caused a single case of sensitization.[20] Recently, two patients sensitized to 4,4'-thiobis(6-*tert*-butyl-*meta*-cresol) in NRL examination gloves have been described.[28] In addition to NRL materials, this chemical can be used in Neoprene® latex and in PE materials. One of the patients was also allergic to butylhydroxyanisole (BHA), used in some brands of NRL gloves. A list of examination gloves containing 4,4'-thiobis(6-*tert*-butyl-*meta*-cresol) and BHA can be found in Reference 28.

The monobenzyl ether of hydroquinone has been used as an antioxidant in rubber gloves.[24,27] Besides being a sensitizer, it was the cause of the first reported case of occupational rubber glove induced leukoderma.[74]

Other potential sensitizers include vulcanizers, e.g., 4,4'-dithiomorpholine and benzoyl peroxide, but sensitization from gloves has not as yet been reported.[3,4] Organic pigments, used both in rubber and plastic materials, may also be leachable and induce sensitization.[75] Chromium[76] and, recently, cetylpyridinium chloride[77] are also suggested to be causes of rubber glove allergy. Apart from additives known to induce contact sensitization, many other chemicals are used in the rubber industry;[23,27] these can possibly be used in the manufacture of gloves, and may prove to be sensitizers.

A case of sensitization to glove powder has been reported, but the allergen was not specified.[78] Preservatives (e.g., sorbic acid, isothiazolin-3-one derivative) and epichlorohydrin are potential sensitizers found in glove powders,[3,4] as well as fragrances used occasionally both in rubber[27] and in plastic gloves.

B. Plastic Gloves

Most allergy problems are connected with the use of PVC gloves. A vinyl glove material may, for example, be composed of about 50% PVC and 50% additives. These include plasticizers, stabilizers, UV absorbers, fungicides, bactericides, flame retardants, and colorants.[20] Although plastic materials are generally considered as nonsensitizers, some additives may be leachable and cause contact sensitization.[2,75] PVC itself is probably not a sensitizer.[33] The few reported cases of allergy to plastic gloves, however, have been unable in most cases to determine the actual sensitizer.[9,10,29,30,34,35] The commonly used patch test series of plastics and glues contain only a few potential allergens in PVC materials, and the ingredients of the suspected glove materials are seldom available for testing. It is therefore understandable that a specific allergen has been detected only in some cases. Sensitization to epoxy resin used as a plasticizer has been reported by Fregert and Rorsman.[32] Epoxy resins can be used as plasticizers and stabilizers in PVC (e.g., gloves), PVAC plastics, and Neoprene®.[31,32] Plasticizers were possibly the cause of one patient's dermatitis, when five patients were sensitized to their PVC gloves. The patient reacted to tricresyl phosphate and triphenyl phosphate, known to be used as plasticizers in PVC.[2] Other potential plasticizer-sensitizers include phthalates, e.g., dibutyl and dioctyl phthalates.[20] In addition, colorants may be the cause of PVC glove dermatitis. The colorant, Irgalite Orange F2G® (CI Pigment Orange 34), was found to be the actual sensitizer in one of the five above-mentioned patients.[9,33,75]

Apart from the above-mentioned agents, there may still be other potential allergens in plastic gloves. The semantics between rubber, especially synthetic rubber, and plastic polymers is somewhat artificial. Therefore, the manufacturers and distributors of gloves may sometimes call a certain material a plastic, whereas others may call a similar material a synthetic rubber. For instance, in Finland a certain brand of gloves made of nitrile rubber polymer or of mixed material has been sold as plastic gloves.[78] On the other hand, some of the same chemicals can be used in both plastics and rubbers, e.g., a certain thiourea derivative.[20,23,79]

IV. CLINICAL ASPECTS

A. Predisposing Factors

The development of type IV contact allergy to polymer gloves is influenced by many factors, such as the quality of the polymers, the amounts and types of substances added to the polymer mixtures, the time of contact with the skin, the quality of the materials handled with gloves, and the health state of the skin of the hands.[78]

The use of rubber gloves, as such, obviously involves more risk of sensitization than the use of plastic gloves, since rubber usually contains more potent and more leachable allergenic additives than plastic. There are also great differences in the materials of rubber gloves, despite the same nominal quality. The composition of the gloves may vary from one manufacturer to another because different chemicals in different amounts are used for the same purposes, e.g., as accelerators and antioxidants. Accordingly, it has been shown that patients allergic to a certain dithiocarbamate have, on patch testing, been positive to some of the gloves containing that particular dithiocarbamate, while remaining negative to some others although they contain the same chemical.[56,63]

Longstanding daily use of rubber gloves, simultaneous exposure of hands to chemical or mechanical irritants, or wet work seem to constitute the most important risk of developing allergy to glove materials. Probably, women at work wear rubber (and plastic) gloves more often than men. They are also more often employed in jobs which require hand protection, and the use of gloves is also common in households. Allergy to gloves seems, accordingly, to be more frequent among women.[9,10,16,20,22,38,39,80,81] However, allergy to gloves is also common in male workers who regularly use rubber gloves.[4,10,20,21,26,37,39,55,57,70,83,84,90] In addition to housewives, risk occupations include work in the medical and dental health services (surgeons, gynecologists, orthopedists, nurses, instrument caretakers, dentists, dental nurses), hospital and other cleaners, hairdressers, kitchen workers or workers in food industries, farmers, and industrial workers, e.g., masons and construction workers.

It is not completely known how additives are leached from the vulcanized products, but perspiration and high temperature may promote the release of these agents[20,23,56] from the products, and thus enhance the development of allergy. Degradation of rubber by heat, friction, and ozone may also contribute to the development of sensitization.[23] Experimental studies in conditions simulating normal use of gloves have even proved the release of additives from gloves. At the same time, some degradation of the released chemicals has been found to take place.[56]

Allergens easily penetrate the skin if it is macerated or inflamed. Irritant dermatitis or allergic contact dermatitis from other reasons often precedes sensitization to rubber. Often, the use of gloves has been started only after the appearance of skin irritation or hand eczema.[9,16,21,37] Sensitivity to rubber may also mimic or complicate other types of dermatoses such as photodermatitis, pompholyx, psoriasis, or dermatophytosis.[85]

Chemicals coming into contact with the glove materials may also promote the development of allergy to glove ingredients. Organic solvents can extract allergenic compounds from glove materials and carry them inside the glove.[59,86] Some other chemicals, e.g., bleaching agents, can transform the structure of a chemical in rubber products so that it becomes allergenic.[87] Surgeons may treat the thumb, index, and middle fingers of their surgical gloves with sodium hypochlorite to produce good traction of the material,[88] at the same time possibly increasing the risk to be sensitized to the material.

B. Sites and Appearance of Dermatitis

Polymer gloves cause skin symptoms usually on the backs of the hands, but sometimes also on the palms near the thumbs, on both sides of the wrists, and on the forearms. The border of dermatitis on the mid-arm, corresponding with the upper border of the glove, is often abrupt.[20] The dermatitis may first appear as a few small papules on the fingers and then extend to the whole dorsal area of the hands. The symptoms may include swelling, intense redness, vesicles or blisters, scaling or rhagades, itching, and stinging or burning sensations, depending on the stage and duration of the dermatitis (Figure 1). Another typical manifestation is diffuse or patchy eczema on the back of the hands and on the forearms, not extending over the whole glove area.[20] Forearms, face, and also larger areas may be involved in more severe rubber glove dermatitis.[18] Sometimes the face can be more affected than the hands. Two patients sensitized to their rubber gloves and investigated at our clinic had the most pronounced symptoms on their faces, especially the eyelids (Figure 2). Jordan has also described eyelid dermatitis in the absence of hand eczema in patients who are allergic to rubber gloves (cited by Taylor[23]). When rubber allergy superimposes other types of dermatoses, the appearance may not differ much from that of the prevailing dermatosis (see Section IV.A).

Rare manifestations connected with allergy to rubber include leukoderma, and purpuric and lichenoid dermatoses. Depigmented lesions occur rarely nowadays; they may be located on the backs of the hands and arms related to the use of gloves.[24]

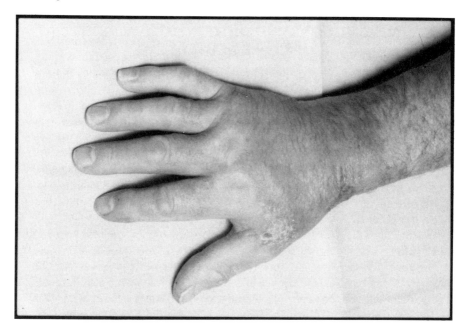

FIGURE 1 Typical appearance of rubber glove dermatitis.

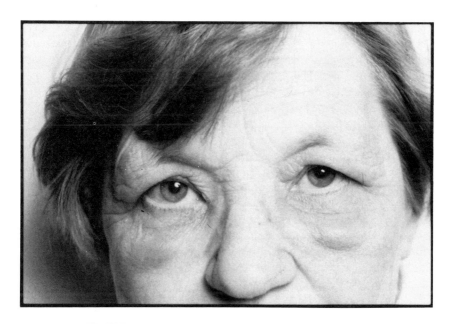

FIGURE 2 Allergy to rubber gloves appearing as eyelid edema.

V. DIAGNOSIS AND DIFFERENTIAL DIAGNOSIS OF GLOVE DERMATITIS

The possibility of glove allergy should be considered in cases where there is recalcitrant dermatitis despite the careful use of rubber or plastic gloves with or without separate inner gloves. The wearing of thin cotton gloves does not always prevent the development of sensitization, as the additives released from gloves may easily soak through the fabric and reach the skin. The possibility of glove sensitization should also be borne in mind in cases where skin problems caused by contact with other rubber objects have been encountered.[20] Also, occasional use of polymer gloves, although not considered important by the patient, must be kept in mind, especially in cases of dermatoses other than contact eczema.

The investigations should include patch testing with pieces of rubber and plastic glove materials, using the 48-h occlusion time. A longer occlusion time has been considered to be necessary, especially when thin materials like surgical gloves are tested.[23] However, the longer occlusion time also increases the risk of false positive reactions. Testing with solid materials always includes a risk of false positive irritant reactions, but retesting and control tests on nonexposed persons, as well as experience in reading the tests,[18] may help to distinguish irritant reactions from allergic ones. Further testing with a standard series, e.g., the European series (Chemotechnique Diagnostics Ab, Malmö, Sweden),[89] is necessary to confirm allergy to rubber. The series contains three important rubber mixes, the thiuram mix, black rubber mix, and carba mix, and MBT which may detect most cases of rubber glove allergy. Particularly, the thiuram mix has been shown to be a good detector of glove allergy.[20,22,63]

The standard series contains no thiourea compounds, but some are included in a rubber additive series, e.g., Chemotechnique Diagnostics Ab.[89] Cross-reactivity occurs between thiourea compounds,[90] but not constantly,[91] therefore each thiourea compound needs to be tested separately. Patch testing with a rubber additive series, e.g., Chemotechnique, which contains 24 chemicals,[89] may increase the accuracy of patch testing and help to find alternative materials for patients having a specific rubber additive allergy. Possible sensitization to rubber polymers themselves, as well as to glove powders, should also be excluded.[3,4] The fragrances used to deodorize the materials may also be causes of glove allergy. A series of plastics and glues containing some plasticizers and stabilizers may help in cases where PVC gloves are suspected. Patch testing with all actual glove components is also important in confirming allergy to polymer glove materials. It would also provide more information on the allergenic compounds found in gloves. Tests with ultrasonic bath extracts[92] of glove materials may be necessary, and help to detect the sensitizer when the glove components are not available for testing.

Investigations should also include tests to detect type I allergy to glove materials or glove powders (prick or scratch chamber tests, determinations of specific IgE antibodies in serum [e.g., RAST] and challenges with suspected gloves). See also Chapters 13 and 17.

Differential diagnosis should exclude nonimmunologic contact urticaria (NICU), e.g., induced by sorbic acid which is used as a preservative in glove powders.[93] Cholinergic and pressure urticaria[3] and dermographism[1] may also be causes of glove-related discomfort. Irritant reactions caused by irritating agents in materials[47] and in powders may also mimic glove allergy. Profuse sweating as a result of longstanding use of tight gloves without interruption often leads to maceration of the skin. Bacterial endotoxins and ethylene oxide retained in the gloves during sterilization may also resemble allergic glove dermatitis. Furthermore, it should be remembered that chemicals permeating the glove materials also cause hand eczema. These chemicals include nickel sulfate, acrylate compounds, epoxy resins and reactive diluents of epoxy resins, hair dyes, and thioglycolates.[1,3,4,94] See also Chapter 15.

VI. PREVENTION OF GLOVE DERMATITIS

The following points are important in the prevention or amelioration of the problems encountered in the use of polymers. The use of rubber or plastic gloves should be started at the same time as the handling of hazardous materials or wet work. All polymer gloves are designed to protect healthy skin. The use of polymer gloves alone does not solve the problems of a diseased skin. The gloves should be used only temporarily to protect inflamed skin. Every patient who has prolonged hand dermatitis needs a detailed dermatological investigation, including skin tests, to detect possible allergies. A detailed job analysis is always necessary before selecting gloves. Only good quality gloves, with the type of materials clearly marked should be selected for use. Two recent cases of unexpected contact urticaria and an anaphylactic reaction evoked by mixed rubber materials not made known by the manufacturer[78,95] demonstrate that distributors as well as consumers are in need of more reliable and accurate information on glove materials. Plastic gloves should be used whenever possible. Separate inner textile gloves possibly reduce risk of sensitization, and prevent maceration of the skin. When rubber gloves are better alternatives, disposable PVC or PE gloves can sometimes be used as inner gloves to increase protection and prevent sensitization to rubber. Glove powders, when used, should not contain any hazardous agents. Detailed instructions on how to use and care for the gloves should be given to the workers. Good glove care and quality control should be provided for in each workplace.[1]

Patients who have type IV allergy to NRL household or industrial rubber gloves can wear gloves made of PVC or other plastic materials, e.g., the Danish multilayered glove (Safety 4 A/S, Lyngby, Denmark),[96] which is especially resistant to many organic solvents, epoxy compounds, and acrylates. In some cases, synthetic rubber materials may also be alternatives. In cases of type IV allergy to surgical and examination gloves, or in cases of problems with glove powders, there is more information available on alternatives. NRL gloves or synthetic rubber gloves can often be selected according to patch test results with standard and rubber additive series, thus avoiding gloves containing the specific allergen(s). There are also many powder-free gloves available. Lists of recommendable surgical and examination gloves can be found in References 3, 4, 28, 66, and 97.

VII. SUMMARY

Allergic contact dermatitis from rubber gloves is common, whereas allergy to plastic gloves is rare. The frequency of allergic contact dermatitis due to rubber gloves seems to be on the increase. Rubber gloves are considered to be the main source of occupational and nonoccupational sensitization to rubber. They are responsible for more than two thirds of occupationally induced cases of rubber allergy. The occupational groups most frequently affected by contact dermatitis due to rubber gloves are cleaners, kitchen and food industry workers, and health care employees. Thiurams, carbamates, and benzothiazoles are accelerators which are the most frequent causes of sensitization due to rubber gloves. Potential allergens in PVC gloves include stabilizers and plasticizers. In order to prevent allergy to polymer gloves, plastic gloves should be primarily used to protect the hands. Every patient with prolonged hand dermatitis should be sent to a detailed dermatological investigation. A detailed job analysis should be made before glove selection, and only good quality gloves, clearly marked as to the material content, should be worn. Cooperation with glove manufacturers is of utmost importance in the prevention of polymer glove allergy and in the development of safer polymer materials for protective gloves.

REFERENCES

1. Estlander, T. and Jolanki R., How to protect the hands, *Dermatol. Clin.*, 6, 105, 1988.
2. Estlander, T., Occupational Skin Disease in Finland. Observations made during 1974 to 1988 at the Institute of Occupational Health, Helsinki, thesis, *Acta Derm. Venereol.*, Suppl. 155, 1990.
3. Heese, A., von Hintzenstern, J., Peters, K.-P., Koch, H. U., and Hornstein, O. P., Allergic and irritant reactions to rubber gloves in medical health services, *J. Am. Acad. Dermatol.*, 25, 831, 1991.
4. Heese, A., von Hintzenstern, J., Peters, K.-P., and Koch, H. U., Typ IV-Allergien gegen Gummihandschuhe — Inzidenz, Allergene, Diagnostik und Therapie. Allergies of the delayed type against rubber gloves. Incidence, allergens, diagnostic approach, therapy, *Z. Hautk.*, 66, 25, 1991.
5. Turjanmaa, K., Latex Glove Contact Urticaria, thesis, University of Tampere, Finland, 1988.
6. Turjanmaa, K. and Reunala, T., Latex-contact urticaria associated with delayed allergy to rubber chemicals, in *Current Topics in Contact Dermatitis,* Frosch, P. J., Dooms-Goossens, A., Lachapelle, J.-M., Rycroft, R. J. G., and Scheper, R. J., Eds., Springer-Verlag, Berlin, 1989, 460.

7. Tarvainen, K., Jolanki, R., and Forsman-Grönhom, L., Reinforced plastics industry: exposure, skin disease and skin protection, in *Proc. Quality and Usage of Protective Clothing*, Mäkinen, H., Ed., Nokobetef IV, Finland, 1992, 147.
8. van Ketel, W. G., Contact urticaria from rubber gloves after dermatitis from thiurams, *Contact Derm.*, 11, 323, 1984.
9. Estlander, T., Jolanki, R., and Kanerva, L., Dermatitis and urticaria from rubber and plastic gloves, *Contact Derm.*, 14, 20, 1986.
10. Frosch, P. J., Born, C. M., and Schutz, R., Kontaktallergien auf Gummi-, Operations- und Vinylhandschuhe, *Hautartz*, 38, 210, 1987.
11. Osmundsen, P. E., Contact urticaria from nickel and plastic additives (butylhydroxytoluene, oleylamide), *Contact Derm.*, 6, 452, 1980.
12. Mitchell, J. M., Contact urticaria from plastic shower curtains, *Contact Derm.*, 9, 329, 1983.
13. Bonnevie, P. and Marcussen, P. V., Rubber products as a widespread cause of eczema. Report of 80 cases. *Acta Derm. Venereol.*, 25, 164, 1944.
14. Pirilä, V., Dermatitis due to rubber, *Acta Derm. Venereol.*, 2, 252, 1957.
15. Schultheiss, E., Gummi und Ekzem, *Berufsdermatosen*, Monographien zur Zeitschrift "Berufsdermatosen", Band 2, Editio Cantor, Aulendorf, 1959.
16. Wilson, H. T. H., Rubber-glove dermatitis. *Br. Med. J.*, 2, 21, 1960.
17. Wilson, H. T. H., Rubber dermatitis. An investigation of 106 cases of contact dermatitis caused by rubber, *Br. J. Dermatol.*, 81, 175, 1969.
18. Nurse, D. S., Rubber sensitivity, *Aust. J. Dermatol.*, 20, 31, 1979.
19. Song, M., Degreef, H., De Maubeuge, J., Dooms-Goossens, A., and Oleffe, J., Contact sensitivity to rubber additives in Belgium, *Dermatologica*, 158, 163, 1979.
20. Cronin, E., *Contact Dermatitis*, Churchill Livingstone, New York, 1980, 714.
21. Themido, R. and Menezes Brandão, F., Contact allergy to thiurams, *Contact Derm.*, 10, 251, 1984.
22. Lammintausta, K. and Kalimo, K., Sensitivity to rubber. A study with rubber mixes and individual rubber chemicals, *Dermatosen*, 33, 204, 1985.
23. Taylor, J. S., Rubber, in *Contact Dermatitis*, 3rd ed., Fisher, A. A., Ed., Lea & Febiger, Philadelphia, 1986, 603.
24. Conde-Salazar, L., Rubber dermatitis, clinical forms, *Dermatol. Clin.*, 8, 49, 1990.
25. Foussereau, J., Tomb, R., and Cavelier, C., Allergic contact dermatitis from safety clothes and individual protective devices, *Dermatol. Clin.*, 8, 127, 1990.
26. Maso, M. J. and Goldberg, D. J., Contact dermatoses from disposable glove use: a review, *J. Am. Acad. Dermatol.*, 23, 733, 1990.
27. Menezes Brandão, F., Rubber, in *Occupational Skin Disease*, 2nd ed., Adams, R. M., Ed., W.B. Saunders, Philadelphia, 1990, 462.
28. Rich, P., Belozer, M. L., Norris, P., and Storrs, F. J., Allergic contact dermatitis to two antioxidants in latex gloves: 4,4'-thiobis(6-*tert*-butyl-*meta*-cresol) (Lowinox 44S36) and butylhydroxyanisole, *J. Am. Acad. Dermatol.*, 24, 37, 1991.
29. Templeton, H. J., Contact dermatitis from plastic mittens, *Arch. Dermatol. Syph.*, 61, 854, 1950.

30. Morris, G. E., Vinyl plastics, *Arch. Ind. Hyg. Occup. Med.,* 8, 535, 1953.
31. Fregert, S. and Rorsman, H., Hypersensitivity to epoxy resins used as plasticizers and stabilizers in polyvinylchloride (PVC) resins, *Acta Dermatol. Venereol.,* 43, 10, 1963.
32. Fregert, S. and Rorsman, H., Allergens in epoxy resins, *Acta Allergol.,* 19, 296, 1964.
33. Kanerva, L., Jolanki, R., and Estlander, T., Organic pigment as a cause of plastic glove dermatitis, *Contact Derm.,* 13, 41, 1985.
34. Guillet, M. H., Ménard, G., and Guillet, G., Sensibilisation de contact aux gants en vinyl, *Ann. Dermatol. Venereol.,* 118, 723, 1991.
35. Krasteva, M., Chefai, M., Dupouy, M., and Chabeau, G., Vinyl glove intolerance, in Book of Abstracts, 1st Congr. Eur. Soc. Contact Dermatitis, Brussels, 8-10 October, 1992, 73.
36. Downing, J. G., Dermatitis from rubber gloves, *N. Engl. J. Med.,* 208, 196, 1933.
37. Fregert, S., Occupational dermatitis in a 10-year material, *Contact Derm.,* 1, 96, 1975.
38. Lachapelle, J.-M. and Tennstedt, D., Epidemiological survey of occupational contact dermatitis of the hands in Belgium, *Contact Derm.,* 5, 244, 1979.
39. Romaguera, C. and Grimalt, F., Statistical and comparative study of 4600 patients tested in Barcelona (1973–1977), *Contact Derm.,* 6, 309, 1980.
40. Kanerva, L., Estlander, T., and Jolanki, R., Occupational skin disease in Finland. An analysis of 10 years of statistics from an occupational dermatology clinic, *Int. Arch. Environ. Health,* 60, 89, 1988.
41. Jolanki, R., Estlander, T., and Toikkanen, J., Ammatti-ihotaudit Suomessa. Havaintoja työperäisten sairauksien rekisteristä. (Occupational skin diseases in Finland. Observations from Finnish Occupational Disease Register), *Työterveyslääkäri,* 3, 18, 1989.
42. Kanerva, L., Jolanki, R., and Toikkanen, J., Työperäiset allergiat v. 1991 (Cases due to occupational allergy in 1991), Institute of Occupational Health, Helsinki, 1992.
43. Fisher, A. A., *Contact Dermatitis,* 3rd ed., Lea & Febiger, Philadelphia, 1986.
44. West, D. F., New trends in rubber and plastic clothing, *Saf. Maint. Prod.,* 103, 18, 1952.
45. Aoyma, M., Sugiura, K., Fujise, H., and Naruse, M., On use of gloves in the home and their influence upon skin irritation, *Nagoya Med.,* 27, 65, 1982.
46. Ishihara, M., Diagnostics of contact dermatitis caused by rubber or vinyl gloves, *J. Therap.,* 51, 166, 1969.
47. Naruse, M. and Iwama, M., Dermatitis from household vinyl gloves, *Bull. Environ. Contam. Toxicol.,* 48, 843, 1992.
48. Mellström, G., Protective Gloves of Polymeric Materials. Experimental Permeation Testing and Clinical Study of Side Effects, thesis, Arbetsmiljöinstitutet, *Arbete och Hälsa,* 10, 1991.
49. Morton, M., *Rubber Technology,* 2nd ed., Van Nostrand Reinhold, New York, 1973.
50. Kortschak, E., Rubber Chemicals, *Aust. J. Dermatol.,* 18, 127, 1977.
51. Lezaun, A., Marcos, C., Martín, J. A., Quirce, S., and Díez Gúmez, M. L., Contact dermatitis from natural latex, *Contact Derm.,* 27, 334, 1992.

52. Wyss, M., Elsner, P., and Wüthrich, B., Allergic contact dermatitis to natural latex, *J. Am. Acad. Dermatol.*, 27, 650, 1992.
53. Adams, R. M., Dermatitis due to clothing, *Cutis*, 5, 577, 1972.
54. Fregert, S., Cross-sensitivity pattern of 2-mercaptobenzothiazole (MBT), *Acta Dermatol. Venereol.*, 49, 45, 1969.
55. Fisher, A. A., Contact dermatitis in surgeons, *J. Dermatol. Surg.*, 1, 63, 1975.
56. Knudsen, B. B., Larsen, E., Egsgaard, H., and Menné, T., Release of thiurams and carbamates from rubber gloves, *Contact Derm.*, 28, 63, 1993.
57. Nater, J. P., Überempfindlichkeit gegen Gummi, *Berufsdermatosen*, 23, 161, 1975.
58. Hansson, C. and Ågrup, G., Stability of the mercaptobenzothiazole compounds, *Contact Derm.*, 28, 29, 1993.
59. Estlander, T., Kilpikari, I., Eskolin, E., Polymeerikäsineen läpäisevyys ja sen valmistuksessa käytettävät kemikaalit (Permeability of polymer gloves and chemicals used in the manufacture of the gloves), *Suom. Lääkäril.*, 35, 800, 1980.
60. Jung, J. H., McLaughlin, J. L., Stannard, J., and Guin, J. D., Isolation, via activity-directed fractionation, of mercaptobenzothiazole and dibenzothiazyl disulfide, 2 allergens responsible for tennis shoe dermatitis, *Contact Derm.*, 19, 254, 1988.
61. Wang, X. and Suskind, R., Comparative studies of the sensitization potential of morpholine, 2-mercaptobenzothiazole and 2 of their derivatives in guinea pigs, *Contact Derm.*, 19, 11, 1988.
62. Guin, J. D., The MBT controversy, *Am. J. Contact Derm.*, 1, 195, 1990.
63. Van Ketel, W. G. and van der Berg, W. H. H. W., The problem of the sensitization to dithiocarbamates in thiuram-allergic patients, *Dermatologica*, 169, 70, 1984.
64. Zugerman, C., Allergy to zinc dimethyldithiocarbamate in rubber gloves, *Contact Derm.*, 7, 337, 1981.
65. Higaki, S, Maruyama, T, Takahashi, S., and Morohashi, M., A case of contact dermatitis due to surgical gloves, *So. Skin Res.*, 9, 32, 1990.
66. Scollhammer, M., Guillet, M. H., and Guillet, G., Les dermatoses de contact aux gants médicaux, *Ann. Dermatol. Venereol.*, 118, 731, 1991.
67. Fregert, S., Dahlquist, I., and Trulsson, L., Sensitization capacity of diphenylthiourea and phenylisothiocyanate, *Contact Derm.*, 9, 87, 1987.
68. Masmoudi, M. L. and Lachapelle, J. M., Occupational dermatitis to dihydroxydiphenyl and diphenylthiourea in Neoprene® gloves, *Contact Derm.*, 16, 290, 1987.
69. Camarasa, J. G., Romaguera, C., Conde-Salazar, L., Pascual, A. M., Moran, M., Alomar, A., and Grimalt, F., Thiourea reactivity in Spain, *Contact Derm.*, 12, 220. 1985.
70. Polano, M. K., Ekzem durch Gummihandsschuhe, *Dermatologica*, 116, 105, 1958.
71. Kaniwa, M.-A., Kojima, S., Nakamura, A., Kanto, H., Ito, M., and Ishihara, M., Chemical approach to contact dermatitis caused by household products. II. Analysis of antioxidants in commercial rubber gloves and incidence of positive reactions to antioxidants in patch testing, *Eisei Kagaku*, 30, 126, 1984.

72. Hervé-Bazin, B., Foussereau, J., and Cavelier, C., L'allergie á la *N*-isopropyl-*N'*-phényl*para*phénylenediamine dans certains objets de protection individuelle, *Dermatosen*, 28, 82, 1980.
73. Kaniwa, M.-A., Kojima, S., Nakamura, A., and Ishihara, M., Chemical approach to contact dermatitis by household products. I. *N*-isopropyl-*N'*phenyl-*p*-phenylenediamine in heavy duty rubber gloves, *Eisei Kagaku*, 28, 137, 1982.
74. Oliver, E. A., Schwartz, L., and Warren, L. H., Occupational leucoderma. A preliminary report, *J. Am. Med. Assoc.*, 113, 927, 1939.
75. Jolanki, R., Kanerva, L., and Estlander, T., Organic pigments in plastics can cause allergic contact dermatitis, *Acta Dermatol. Venereol.*, Suppl. 134, 95, 1987.
76. Conde-Salazar, L., Castano, A. M., and Vicente, A., Determination of chrome in rubber gloves, *Contact Derm.*, 6, 237, 1980.
77. Castelain, M. and Castelain, P.-Y., Allergic contact dermatitis from cetyl pyridinium chloride in latex gloves, *Contact Derm.*, 28, 118, 1993.
78. Milkovic-Kraus, S., Glove powder as a contact allergen. *Contact Derm.*, 26, 198, 1992.
79. Estlander, T., Jolanki, R., and Kanerva, L., Contact urticaria from rubber gloves: a detailed description of four cases, *Acta Dermatol. Venereol.*, Suppl. 134, 98, 1987.
80. Fregert, S., Chemicals used in both rubber and plastics, *Contact Derm. Newsl.*, 9, 204, 1971.
81. Wall, L. M. and Gebauer K. A., Occupational skin disease in Western Australia, *Contact Derm.*, 24, 101, 1991.
82. Guerra, L., Tosti, A., Bardazzi, F., Pigatto, P., Lisi, P., Santucci, R., Valsecchi, D.,Schena, D., Angelini, G., Sertoli, A., Ayala, F., and Kokelj, F., Contact dermatitis in hairdressers: the Italian experience, *Contact Derm.*, 26, 101, 1992.
83. van der Meeren, H. L. M., Boezeman, J. B. M., and Rampen, F. H. J., Contacteczeem door operatiehandschoenen, *Ned. Tijdschr. Geneskd.*, 132, 963, 1988.
84. Jacobsen, N., Aasenden, R., Hensten-Pettersen, A., Occupational health complaints and adverse reactions as perceived by personnel in public dentistry, *Community Dent. Oral Epidemiol.*, 19, 155, 1991.
85. Fisher, A. A., Management of dermatitis due to surgical gloves, *J. Dermatol. Oncol.*, 11, 628, 1985.
86. Williams, J. R., Permeation of glove materials by physiologically harmful chemicals, *Am. Ind. Hyg. Assoc. J.*, 40, 877, 1979.
87. Jordan, W.P. and Bourlas, M. C., Allergic contact dermatitis to underwear elastic chemically transformed by laundry bleach, *Arch. Dermatol.*, 111, 593, 1975.
88. Guin, J. D., The doctor's surgical/examination gloves — problems with and without them, *Int. J. Dermatol.*, 31, 853, 1992.
89. Product Catalogue, Patch Test Allergens, Chemotechnique Diagnostics, Malmö, Sweden, 1992/93.
90. Roberts, J. L. and Hanifin, J., Contact allergy and cross reactivity to substituted thiourea compounds, *Contact Derm.*, 6, 138, 1980.
91. Kanerva, L., Estlander, T., and Jolanki, R., Contact allergic reactions to thioureas, in Book of Abstracts, 18th World Congr. Dermatol., New York, June 12-18, 1992, 109A.

92. Bruze, M., Trulsson, L., and Bendsöe, N., Patch testing with ultrasonic bath extracts, *Am. J. Contact Derm.*, 3, 133, 1992.
93. Lahti, A. and Maibach, H. I., Immediate contact reactions, in *Exogenous Dermatoses: Environmental Dermatitis*, Menné T. and Maibach, H.I., Eds., CRC Press, Boca Raton, FL, 1991, 21.
94. Storrs, F. J. and Portland, O. R., Permanent wave contact dermatitis: contact allergy to glyceryl monothioglycolate, *J. Am. Acad. Dermatol.*, 11, 74, 1984.
95. Heese, A., Peters, K.-P., and Hornstein, O. P., Anaphylactic reaction to unexpected latex in a polychloroprene glove, *Contact Derm.*, 27, 336, 1992.
96. Chemical Protection List, 4 H, Safety 4 A/S Lyngby, Denmark, 1/1992.
97. Lahti, A., Camarasa, J. G., Ducombs, G., Lachapelle, J.-M., Maibach, H. I., Menné, T., Niinimäki, A., Wahlberg, J. E., Wilkinson, J. D., and Wrangsjö, K., Patch tests with Tactylon® in patients with contact allergy to rubber, *Contact Derm.*, 27, 188, 1992.
98. Estlander, T., Jolanki, R., and Kanerva, L., unpublished data.
99. Kanerva, L., Estlander, T., and Jolanski, R., Occupational allergic contact dermatitis caused by thiourea compounds, *Contact Derm.*, in press.

17

Contact Urticaria from Latex Gloves

Kristiina Turjanmaa

TABLE OF CONTENTS

I. Introduction 241
II. Contact Urticaria Syndrome 242
III. Diagnosis 243
    A. Skin Prick Test 243
    B. Use Test 245
    C. RAST 245
IV. Symptoms 245
V. Frequency 247
VI. Risk Factors 248
VII. Concomitant Type IV and Type I Allergy to Rubber Products 249
VIII. Rubber Protein Allergens 249
IX. Prophylactics 250
X. Conclusions 251
References 251

I. INTRODUCTION

Rubber allergy has been known for decades as a cause of eczema among glove users.[1] It has also been a marked occupational disorder among workers in different fields who want to protect their hands from irritative and sensitizing chemicals.[2] Little was known about the more far-reaching consequences of rubber allergy before 1979, when Nutter reported his first case of immediate contact urticaria in hands after the use of household gloves.[3] It soon became evident that contact urticaria from gloves not only causes itching but can even give rise to life-threatening symptoms.[4-6] Förström fixed attention on the occupational nature of this allergy in his case study of a nurse who wore surgical gloves.[7] The "old" rubber allergy represents a type IV sensitization to the chemicals used in rubber manufacture; the "new" one is an immediate, type

TABLE 1 NRL Products Giving Symptoms in 57 Health Care Workers and in 67 Adults of Other Professions With NRL Allergy

Product	Health care workers (n = 57)	Others (n = 67)
Household gloves	14 (25%)	57 (85%)
Surgical gloves	54 (95%)	4 (6%)
Balloons	4 (7%)	18 (27%)
Condoms	9 (16%)	6 (9%)
Rubber bands	3 (5%)	5 (7%)
Medical intervention	5 (9%)	8 (12%)
Aerogen	9 (16%)	5 (7%)

I, IgE-mediated allergy to proteins originating from the rubber tree *(Hevea brasiliensis)* which are still present in products made from natural rubber latex (NRL).[8-10] The symptoms of IgE-mediated allergy extend from contact urticaria and eczema to generalized urticaria, conjunctivitis, rhinitis, asthma, and anaphylaxis; even lethal cases have been described.[11-13] It can be elicited from all NRL products such as pacifiers, balloons, toys, condoms, and medical devices, but surgical and protective gloves are the most important cause of symptoms (Table 1).[14] It is a major occupational hazard among health care workers, but another group of people especially at danger are sensitized but undiagnosed persons who come into contact with different kinds of NRL products as patients in the health care sector.[15-17] Because of the complexity of the problems the U.S. Food and Drug Administration organized a three-day conference dealing only with allergy to NRL in medical devices in November 1992.

II. CONTACT URTICARIA SYNDROME

Maibach and Johnson defined contact urticaria (CU) as a wheal-and-flare response to the application of chemicals to intact skin.[18] It usually appears within a few to 60 min after contact and disappears within some hours, at the latest in 24 h. The two main types are nonimmunologic and immunologic CU, although in some cases the mechanism remains unknown. Immunologic CU is caused by proteins or protein complexes, and previous sensitization is needed for elicitation of symptoms. The term contact urticaria syndrome (CUS) has been used for immunologic CU cases in whom the urticarial response is not confined to the contact area. It has been divided into four clinical stages by Maibach and Johnson:[18] stage 1, localized urticaria in the contact area; stage 2, generalized urticaria with angioedema; stage 3, urticaria with bronchial asthma as well as rhinoconjunctivitis, orolaryngeal, and gastrointestinal symptoms; and stage 4, urticaria with anaphylactic reaction. The localized reaction can also appear in the form of tiny itchy vesicles leading to eczema or to the maintenance of a preexisting eczema when the protein allergen comes into contact with the skin repeatedly. The exact mechanism whereby the large protein molecules penetrate the epidermis is not known, but once there, the allergens react with the IgE receptors on mast cell membranes, leading to the release of vasoactive substances, mainly histamine, and the eruption of wheal-and-flare. There are also IgE receptors on Langerhans cells. The binding of the allergen with these elicits a similar reaction to that which occurs in delayed-type contact dermatitis, resulting in "protein contact dermatitis".[19]

Natural rubber latex (NRL) allergy is a typical example of immunologic contact urticaria, with a positive Prausnitz-Küstner test verifying its IgE-mediated mechanism.[8,9] The distribution of the symptoms elicited by NRL gloves was analyzed in a study comprising 42 subjects: no local skin symptoms were seen in 2%; redness, itching or exacerbation of hand eczema in 10%; CUS stage 1 in 40%; stage 2 in 17%; stage 3 in 21%; and stage 4 in 10%.[11]

Division of the symptoms of CU into four different stages seems to be clinically justified. In the case of allergy to NRL gloves, conjunctivitis might be classified as stage 1, since the symptoms often appear after touching the eye with an ungloved hand after gloves have been worn, and the symptom is consequently one-sided. Rhinitis may also be caused by direct mucosal contact with airborne NRL-contaminated particles of the corn starch powder used in the manufacture of latex gloves. Gastrointestinal symptoms may also be related to direct mucosal contact with allergens eluted from gloves when eating food, e.g., sandwiches prepared by a person wearing gloves. Thus far, there are no publications confirming these hypotheses.

III. DIAGNOSIS

Different methods and different allergens have been used for diagnosing the immediate allergy to NRL.[3,7-9,20-23] Because no 100% sensitive and specific allergens are available commercially, each team of investigators has developed its own. This makes definition difficult and the comparison of relevancy between different papers impossible. Questionnaires do not even reveal the persons at high risk.[24,25] Skin prick testing with glove eluates has proved to be a reasonably reliable and safe method as long as certain criteria are followed.[11,26] In doubtful cases, the diagnosis can be verified with a use test with glove material. The latex RAST is useful, but at this moment its sensitivity is still too poor for it to be used as a screening method.[15] The latex histamine release test has good sensitivity and specificity, but because of the complicated methodology it is not suitable for the study of large groups of patients.[27]

A. Skin Prick Test

Glove material and natural rubber latex, with or without ammonia as a preservative, have been used as allergens. The choice of methods is considerable: open patch test, rub test, scratch and scratch-chamber test, and skin prick test (SPT).[7-9,20-23] Several types of lancets can be used for the skin prick or puncture tests. Needles can also be used, but the results are not always comparable.[28] The use of excessively strong allergens can lead to generalized side effects, which have frightened investigators not used to performing type I skin tests.[10] Lacking consensus of methodology can also be misleading: prick (puncture) test and intradermal (intracutaneous) test methods should not be intermixed; the latter should never be used for testing strong allergens like NRL.[29,30]

So far, only one comparison of different diagnostic methods of NRL allergy has been published.[31] SPT reactions using a homemade glove eluate and a liquid from the rubber tree, and a scratch-chamber test using crushed leaf of the rubber tree elicited concordant results in 13/15 patients verified as having the immediate allergy to NRL gloves by use test and positive SPT reactions to several other brands of latex gloves. The latex glove allergen is made as follows: 1 g of small pieces of a latex glove is incubated in 5 ml of sterile physiological saline taken from a container without previous rubber contact. After 15 min the glove pieces are removed, and the solution is stored in a sterile bottle and used for SPT. No preservatives are needed if the test material is kept at 4°C and a new solution is prepared once a month. Glove brands vary greatly in allergenicity, and therefore a known allergenic brand should be chosen.[32] It is recommended that three different brands be tested at the same time, because the quality of one brand may be changed by the manufacturer without warning. In a study of 100 patients, 99 showed positive SPT with glove eluates.[33] The only SPT-negative person had a positive use test and a positive histamine liberation test with NRL allergen.

The use of the gloves as test material has been criticized because they may also contain other water-soluble elements causing type I reactions in the skin.[34] In our laboratory, 80 patients with NRL allergy have now been retested concomitantly with glove eluates and dilutions of nonammoniated, nonthiuram-containing rubber tree liquid. The results are 100% concordant. When low-ammonia-tetramethylthiuram-disulfide-zinc oxide (LATZ) is used for SPT, it is possible to get reactions from thiurams. Type I reactions to rubber chemicals are still questionable, because the test materials are in petrolatum, and the solubility of rubber chemicals from that substance is badly controlled.[35] The test reactions have not always been reproducible, and in some case reports the possibility of type I allergy to NRL has not been fully excluded.[36]

Commercial NRL allergens are now available: the material of ALK is used in the Scandinavian countries, that of Stallergenes in France and possibly elsewhere, and of Bencard in Canada.[15,37] Their specificity, sensitivity, and safety have not yet been verified by large numbers of patients, but results from Canada and France are promising.

The use of a 1-mm lancet for SPT and different glove eluates as allergens according to the guidelines of the Sub-Committee on Skin Tests of the European Academy of Allergology and Clinical Immunology has never caused any systemic side effects in 142 patients with type I allergy to NRL at our clinic, even when babies and patients with previous anaphylactic reactions caused by NRL have been tested and several glove brands have been used concomitantly.

B. Use Test

The use test with a rubber latex glove should always be performed when there is some discrepancy between the SPT result and clinical history. It is not recommended for patients with a history of NRL-related anaphylaxis when SPT and radioallergosorbent tests (RAST) are positive.

Because the whole-hand use test on eczematous skin has been observed to cause anaphylaxis, the use test should always be started with one finger of a glove on a wet finger.[14,38] If there are some wheals after 15 min, the test is judged as positive and stopped. If not, the test is continued with the whole glove on the wetted hand for another 15 min, always keeping a vinyl glove on the other hand as a negative control. When these guidelines are followed there have been no more adverse reactions to the use test. It often happens that the patient is convinced about a latex glove allergy, and if the 15-min whole-glove use test remains negative, a prolonged-use test is recommended. For up to a week the patient wears a latex glove on one hand and a vinyl glove on the other in daily work. So far, one patient developed CU after 4 h; in all other cases the prolonged-use test has been negative, confirming that not even protein contact dermatitis is caused by the latex gloves.[31]

Patients with NRL-related rhinitis should be challenged on nasal mucosa if the verification is needed for insurance or other authorities in occupational cases. The same glove extract can be used as for SPT. Jaeger et al. have published the guidelines for pulmonal challenge.[39]

C. RAST

The radioallergosorbent test of Pharmacia (Sweden) has long been the only commercially available diagnostic method for NRL allergy. Its specificity is high but its sensitivity is only 60 to 65%.[11,15] In patients with previous anaphylactic reactions the sensitivity is higher, but still not 100%. Homemade disks have proved to be better, giving positive results up to 86%.[40] The sensitivity of the new Pharmacia latex CAP-RAST is expected to be better in the near future. The numerical value of RAST does not correlate with the severity of symptoms elicited by NRL, as shown in Table 2.

IV. SYMPTOMS

The typical symptom from type I allergy to latex gloves is contact urticaria, i.e., wheal-and-flare reaction at the site of contact.[11] The thinner skin of the dorsal side of hands and fingers reacts more easily, as do the wrists. The symptoms begin in a few to 30 min after contact and disappear without treatment in 1 to 2 h. Some people experience only itching and redness, some develop eczema. Because eczema on hands often looks the same, whether caused by irritation, type IV allergy to rubber chemicals, or type I protein contact dermatitis from latex gloves, it always needs to be examined for all these possibilities.[41]

TABLE 2 RAST[a] Results in 11 Patients With Previous Anaphylactic Reaction in Connection With Surgical Technique

Patient no.	Operation	RAST (kU/l)	Total IgE (kU/l)
1.	Delivery, episiotomy	18.3	2867
2.	Delivery, episiotomy	1.2	55
3.	Delivery, episiotomy	0.5	641
4.	Cesarean section	1.3	158
5.	Cesarean section	6.9	1090
6.	Cesarean section	0.81	239
7.	Vaginal ultrasonography (covered by condom)	2.7	95
8.	Use test with gloves	3.3	4388
9.	Measurement of esophageal pressure	0.64	25
10.	Operation of varices venarum	4.3	3380
11.	Operation of varices venarum	1.2	25

[a] Pharmacia Diagnostics, Sweden.

With regard to CU from gloves, there are always two possible types of sensitized recipients: those wearing the gloves and those being touched with gloves worn by others, e.g., nursing personnel. Both can contract the whole variety of CUs described above. The different symptoms are described in Table 3. The 124 patients presented there have been diagnosed in our laboratory since 1984.[33,42]

In Finland, dermatologists perform both prick and patch testing, and most patients with hand eczema are tested with both methods. Thus, our study differs from most others, in which health care workers and children with spina bifida predominate. In a recent paper by Baur et al. about 90% of the patients originated from the health care sector.[43] Most of the patients in our unselected allergologic and dermatologic files have a nonmedical profession (Table 4). This suggests that by examining all atopic patients with a routine series including NRL allergen, and by also using it for every hand eczema patient, we can diagnose far more patients with NRL allergy than previously (Table 5).

TABLE 3 Symptoms Caused by NRL Products in 57 Health Care Workers and in 67 Adults of Other Professions With Verified Immediate Allergy to NRL

Symptom	Health care workers (n = 57)	Others (n = 67)
Contact urticaria	79%	72%
Hand eczema	42%	64%
Conjunctivitis	28%	16%
Rhinitis	16%	13%
Facial edema	14%	28%
Asthma	2%	4%
Generalized urticaria	9%	13%
Anaphylaxis	7%	10%

TABLE 4 Professions of 49/57 (86%) Health Care Workers and of 17/67 (25%) Nonmedical Workers With Work-Related Allergy to NRL

Health care workers		Other occupations	
Physicians	14	Farmers wives'	2
Nurses	17	Kitchen workers	3
Dental nurses	6	Cleaners	3
Laboratory nurses	4	Textile workers	2
Physiotherapists	1	Workers in rubber band plant	3
Workers in a medical plant	3	Paper mill worker	1
Other medical occupations	4	Dairy workers	1
		Departmental secretary	1
		Private caretaker	1

TABLE 5 Cause of Referral to Allergy Testing in 124 Adult Patients With Verified NRL Allergy

Symptom	Health care workers (n = 57)	Others (n = 67)
Hand eczema	27 (47%)	22 (33%)
Atopic eczema	12 (21%)	26 (39%)
Intraoperative anaphylaxis	3 (5%)	7 (10%)
Routine hospital screening	15 (26%)	0
Facial eczema	0	4 (6%)
Rhinitis	0	5 (7%)
Urticaria	0	2 (3%)
Other dermatoses	0	1 (2%)

Women predominate in all published papers dealing with allergy from rubber gloves. This is thought to be because they wear gloves more frequently than men. In the health care sector women predominate among nurses, but among doctors the sexual bias is not so clear, and thus in our material there are ten female and six male doctors who are sensitized.

V. FREQUENCY

Since the first report of Nutter in 1979, there have been several hundreds of case reports on immediate allergy to NRL, mostly in connection with the use of gloves.[11,14,15,44-47] The incidence seems to be increasing, but many of the diagnosed cases have had symptoms for years — for up to 20 years in some cohorts.[48] It may be only the increasing awareness and the better diagnostic facilities that reflect the rising number of diagnosed cases. On the other hand, the use of both surgical latex gloves and, particularly, of examination gloves has increased immensely since the outbreak of hepatitis and HIV infection.

Exposure to other NRL products such as protective gloves, balloons, and condoms is also increasing. There are, however, no exact numbers on the incidence and prevalence of the NRL allergy. In different cohorts of operating room personnel, 2.5 to 10.7% of the tested workers have been diagnosed as having immediate allergy to NRL.[25,48,49] In a small cohort (n = 77) tested by Cormio et al., 99% of the workers in an operating room were skin prick tested with three different latex glove eluates.[50] All solitary weak positive reactions (in size at least half of that of histamine dihydrochloride 10 mg/ml) were controlled by a use test. They all turned out to be false SPT positive: all suspected persons wore without symptoms the same glove brand for 90 min as that which had given the weak positive SPT reaction. All four (5.2%) persons with at least histamine-size reactions to two or three different glove brands showed relevant immediate NRL glove allergy. These results demonstrate the importance of the use test in judging the relevancy of weak skin test reactions.

The Pirkanmaa Health Care District in connection with Tampere University Hospital covers 430,000 people, and to date a total of 142 people have been diagnosed as having NRL allergy at both our hospital and the largest private allergy laboratory in the town. The current minimum prevalence is thus 33/100,000 inhabitants. Because many patients have managed for years without their allergy being diagnosed, and there are newly sensitized patients, too, this number will grow in the future.

In 1991, 1407 new work-related skin diseases were reported to the Finnish Register for Occupational Diseases;[51] 135 of those presented in health care workers: rubber gloves caused immediate allergy in 16 (12%) of them, and delayed-type allergy to rubber chemicals was seen in another 16 subjects. These numbers are probably too low because health care workers are known to avoid registers of this kind.

VI. RISK FACTORS

In various studies atopy and hand eczema have frequently been connected with type I allergy to NRL.[11,14,15] In a study of the 124 adult persons presented above, 75% of the health care workers and 81% of the people in other occupations had a personal history of atopy and, in the same frequency, hand eczema, but not always in the same person at the same time. Rysted's studies confirm that chronic hand eczema is often a complication of atopic dermatitis, so the concomitant occurrence of the two risk factors is easy to explain.[52] Also, exposure to rubber gloves is enhanced in patients with hand eczema and dry skin because they try to protect their skin. Spina bifida patients, as well as other patient populations frequently operated on and using catheters on a regular basis, constitute another big group of risk persons.[53] The prevalence of NRL allergy among patients to be operated on is not known at present, and therefore it is difficult to state categorically whether preoperative testing should be carried out on a routine basis or not.

VII. CONCOMITANT TYPE IV AND TYPE I ALLERGY TO RUBBER PRODUCTS

Concomitant sensitization both to rubber chemicals and to NRL has been reported. In an earlier study, 1 of 18 patients with delayed type allergy to rubber chemicals showed concomitant sensitization to NRL and, vice versa, of 35 patients with NRL allergy, 5 showed positive patch tests to thiurams.[35]

In the present study, 83 of 124 patients were patch tested either with the European Contact Dermatitis Research Group standard series including different rubber mixes only, or also with a rubber series including individual rubber chemicals. Of these patients, 11/83 (13%) showed positive patch test reactions to some rubber chemical or mixes, 7/11 to thiurams only, and 2/11 to both thiurams and carbamates.

Type IV allergy to NRL has also been reported recently, but the patch tests were performed with LATZ containing thiurams. Consequently, thiuram allergy as a cause of the reaction cannot be excluded without retesting with pure material.[54] Heese et al.[34] has pointed to the possible allergies caused by other ingredients included in rubber gloves, but so far the only verified type I reaction to glove ingredients is that caused by ethylene oxide, which was earlier used for sterilization of latex gloves.[34,55] In one patient, immediate allergy to ZDC (zinc diethyldithiocarbamate) was diagnosed, but at the time NRL allergy was not well known, so even in that case the cause of the allergy was probably sensitization to NRL.[36]

Most glove users think they are allergic to corn starch powder. There are papers dealing with corn starch allergy, but sensitization to NRL has not been properly excluded in most cases.[56-59] The only possible agent in corn starch capable of causing CU would be maize protein, but none of the examined patients showed specific IgE to maize. In one study, 42 patients with immediate allergy to NRL were prick tested with two different glove powders supplied by the manufacturer, without previous contact to gloves, and no SPT reactions were seen.[11] Immunoblotting glove powder taken from one latex glove showed the same allergens as were present in latex gloves and in rubber latex, confirming the contamination of glove powder with NRL allergens.[60] This is clinically relevant since NRL-sensitized patients may develop rhinitis and asthma when opening glove packages and in rooms where glove powder is airborne.[22] In our experiment, some patients had a positive use test with latex glove without verification of the responsible allergen, but no type I allergy to NRL, maize, rubber chemicals, or ethylene oxide could be found.

VIII. RUBBER PROTEIN ALLERGENS

Proteins eluting from different glove brands have been analyzed by high power liquid chromatography (HPLC).[32] The allergenic fractions had distinct peaks at 2000, 5000, and 30,000 Da. Skin prick tests with the different fractions were also performed in sensitized patients with positive results. Immunoblotting studies have recently been used to characterize the allergens in NRL.[39,61-66] The patient sera used by different investigators bind heterogeneously to different proteins ranging from 10 to 100 kDa. Sera from three physicians and two nurses with immediate allergy to latex gloves detected ten allergenic proteins in NRL, nine of them were present in ammoniated latex and two in one latex examination glove brand (Exona, Semperit, Austria).[63] Two more allergenic proteins were found in the glove material, suggesting that during manufacture the proteins may change. Immunoelectrophoretic studies revealed at least ten common antigens in surgical latex gloves, natural rubber latex, and ammoniated latex. In crossed-radioimmunoelectrophoresis six of the ten protein antigens in the surgical glove extract and natural rubber latex bound IgE antibodies from NRL-allergic patients' sera. In concordance with the immunoblotting results, also with this method, one allergen not present in natural rubber latex could be demonstrated.[67]

Recent investigations have revealed common allergens in fruit such as banana and avocado and NRL.[68,69] It has not yet been satisfactorily established whether babies showing positive SPR reactions to banana and NRL at the same time, but without any symptoms from using pacifiers, are really sensitized to NRL or is it only because they are allergic to bananas.[70,71]

IX. PROPHYLACTICS

The best prophylactic method is to eliminate the allergens from latex gloves. The proteins, however, are needed to some extent in the manufacture of gloves. Comparison of the allergenicity of different latex gloves by the SPT method showed great variation, suggesting that it is possible to manufacture rubber latex gloves with such small amounts of allergens that even most of the sensitized persons can use them.[32,48] At the moment we lack a reliable method to estimate *in vitro* allergenicity of NRL products, but once we have it, the latex gloves should be labeled accordingly.

The use of low-allergenic gloves will probably diminish the number of sensitized people, assuming that the allergen content of all other NRL products is lowered to the same extent. PVC gloves are not good for people working in the health care sector, where the larger number of pinholes is not acceptable because of the risk of infections. There are already some sterile surgical nonlatex gloves available (Dermapren®, Ansell International, Melbourne; Neolon®, Deseret Medical Inc., Sandy, Utah; Tactylon™, SmartPractice, Phoenix, AZ; Elastyren®, Danpren Gloves A/S, Albertslund, Denmark), which are used when handling sensitized patients and when the health care worker does not tolerate even the less sensitizing latex gloves. In operating rooms, low-allergenic gloves should be used since the airborne NRL-contaminated glove powder causes conjunctivitis, rhinitis, and even asthma in sensitized workers and in sensitized patients.[15,22] It is not necessary to use unpowdered gloves, because even if the gloves contain minimal amounts of allergens, they are not supposed to contaminate the glove powder.

X. CONCLUSIONS

Contact urticaria from rubber latex gloves is an increasing problem both in the health care sector and among people using different kinds of protecting gloves, not to forget housewives using household gloves. Because the symptoms are variable, the sensitized persons consult doctors from different specialties. Besides allergologists, all doctors, but at least dermatologists, otorhinolaryngologists, pulmonists, gynecologists, and surgeons should be aware of the symptoms, diagnostic possibilities, and prophylactics. All health care workers and people using protecting gloves should be increasingly informed. Accurate information from large populations is difficult to obtain, therefore the manufacturers of rubber latex gloves and other NRL products should produce better quality as regards type I allergenicity, and their products should bear adequate labeling. In addition to information concerning allergenic proteins in NRL products, the rubber chemicals used should be indicated on the label.

REFERENCES

1. Downing, J. G., Dermatitis from rubber gloves, *N. Engl. J. Med.,* 208, 196, 1933.
2. Cronin, E., *Contact Dermatitis,* Churchill Livingstone, Edinburgh, 1980, 714.
3. Nutter, A. F., Contact urticaria to rubber, *Br. J. Dermatol.,* 101, 597, 1979.
4. Axelsson, I. G. K., Eriksson, M., and Wrangsjö, K., Anaphylaxis and angioedema due to rubber allergy in children, *Acta Pediatr. Scand.,* 77, 314, 1988.
5. Turjanmaa, K., Reunala, T., Tuimala, R., and Kärkkäinen, T., Allergy to latex gloves: unusual complication during delivery, *Br. Med. J.,* 297, 1029, 1988.
6. Pecquet, C., Leynadier, F., and Dry, J., Contact urticaria and anaphylaxis to natural latex, *J. Am. Acad. Dermatol.,* 22, 631, 1990.
7. Förström, L., Contact urticaria from latex surgical gloves, *Contact Derm.,* 6, 33, 1980.
8. Köpman, A. and Hannuksela, M., Contact urticaria to rubber, *Duodecim,* 3, 39, 1983.
9. Turjanmaa, K., Reunala, T., Tuimala, R., and Kärkkäinen, T., Severe IgE-mediated allergy to surgical gloves, *Allergy,* 39 (Abstr.), 35, 1984.
10. Frosch, P. J., Wahl, R., Bahmer, F. A., and Maasch, H. J., Contact urticaria to rubber gloves is IgE-mediated, *Contact Derm.,* 14, 241, 1986.
11. Turjanmaa, K., Latex Glove Contact Urticaria, thesis, *Acta Universitatis Tamperensis,* Ser A, Vol. 254, University of Tampere, Finland, 1988.
12. Feczko, P. J., Simms, S. M., and Bakirci, N., Fatal hypersensitivity reaction during a barium enema, *Am. J. Roentgenol.,* 153, 275, 1989.
13. Ownby, D. R., Tomlanovich, M., Sammons, N., and McCullough, J., Anaphylaxis associated with latex allergy during barium enema examinations, *Am. J. Roentgenol.,* 156, 903, 1991.

14. Wrangsjö, K., Wahlberg, J. E., and Axelsson, I. G. K., IgE-mediated allergy to natural rubber in 30 patients with contact urticaria, *Contact Derm.*, 19, 264, 1988.
15. Levy, D. A., Charpin, D., Pecquet, C., Leynadier, F., and Vervloet, D., Allergy to latex, *Allergy*, 47, 579, 1992.
16. Fisher, A. A., Allergic contact reactions in health care personnel, *J. Allergy Clin. Immunol.*, 90, 729, 1992.
17. Tomazic, V. J., Withrow, T. J., Fisher, B. J., and Dillard, S. F., Latex-associated allergies and anaphylactic reactions, *Clin. Immunol. Immunopathol.*, 64, 89, 1992.
18. Maibach, H. I. and Johnson, H. L., Contact urticaria syndrome, *Arch Dermatol.*, 111, 726, 1975.
19. Bruynzeel-Koomen, C., van Wichen, D. F., Toonstra, J., Berrens, L., and Bruynzeel, P. L. B., The presence of IgE molecules on epidermal Langerhans cells in patients with atopic dermatitis, *Arch. Dermatol. Res.*, 278, 199, 1986.
20. Kleinhans, D., Contact urticaria to rubber gloves, *Contact Derm.*, 10, 124, 1984.
21. Wrangsjö, K., Mellström, G., and Axelsson, G., Discomfort from rubber gloves indicating contact urticaria, *Contact Derm.*, 15, 79, 1986.
22. Carrillo, T., Cuevas, M., Munoz, T., Hinojosa, M., and Moneo, I., Contact urticaria and rhinitis from latex surgical gloves, *Contact Derm.*, 15, 69, 1986.
23. Estlander, T., Jolanki, R., and Kanerva, L., Dermatitis and urticaria from rubber and plastic gloves, *Contact Derm.*, 14, 20, 1986.
24. Slater, J. E., Allergic reactions to natural rubber, *Ann. Allergy*, 68, 203, 1992.
25. Lagier, F., Vervloet, D., Lhermet, I., Poyen, D., and Charpin, D., Prevalence of latex allergy in operating room nurses, *J. Allergy Clin. Immunol.*, 3, 319, 1992.
26. Dreborg, S., Skin tests used in type I allergy testing. Position paper, *Allergy*, 44 (Suppl.), 10, 1989.
27. Turjanmaa, K., Räsänen, L., Lehto, M., Mäkinen-Kiljunen, S., and Reunala, T., Basophil histamine release and lymphocyte proliferation tests in latex contact urticaria, *Allergy*, 44, 181, 1989.
28. Engler, D. B., DeJarnatt, A. C., Sim, T. C., Lee, J. L., and Grant, J. A., Comparison of the sensitivity and precision of four skin test devices, *J. Allergy Clin. Immunol.*, 90, 985, 1992.
29. Bonnekoh, B. and Merk, H. F., Safety of latex prick skin testing in allergic patients, *J. Am. Med. Assoc.*, 267, 2603, 1992.
30. Sussman, G. L., In reply to Bonnekoh. *J. Am. Med. Assoc.*, 267, 2603, 1992.
31. Turjanmaa, K., Reunala, T., and Räsänen, L., Comparison of diagnostic methods in latex surgical glove contact urticaria, *Contact Derm.*, 19, 241, 1988.
32. Turjanmaa, K., Laurila, K., Mäkinen-Kiljunen, S., and Reunala, T., Rubber contact urticaria. Allergenic properties of 19 brands of latex gloves, *Contact Derm.*, 19, 362, 1988.
33. Turjanmaa, K., Allergy to rubber proteins. Risk factors in 100 sensitized patients, Book of Abstracts, 237 A, 18th World Congr. Dermatol., New York, 1992.
34. Heese, A., von Hintzenstern, J., Peters, K., Koch, H U., and Hornstein, O P., Allergic and irritant reactions to rubber gloves in medical health services, *J. Am. Acad. Dermatol.*, 25, 831, 1991.

35. Turjanmaa, K. and Reunala, T., Latex contact urticaria associated with delayed allergy to rubber chemicals, in *Current Topics in Contact Dermatitis*, Frosch, P. J., Dooms-Goossens, A., Lachapelle, J.-M., Rycroft, R. J. G., and Scheper, R. J., Eds., Springer-Verlag, Heidelberg, 1989, 460.
36. Helander, I. and Mäkelä, A., Contact urticaria to zinc diethyldithiocarbamate (ZDC), *Contact Derm.*, 9, 327, 1983.
37. Sussman, G.L., Tarlo, S., and Dolovich, J., The spectrum of IgE-mediated responses to latex, *J. Am. Med. Assoc.*, 265, 2844, 1991.
38. Turjanmaa, K. and Reunala, T., Contact urticaria from rubber gloves, *Dermatol. Clin.*, 6, 47, 1988.
39. Jaeger, D., Kleinhans, D., Czuppon, A. B., and Baur X., Latex-specific proteins causing immediate-type cutaneous, nasal, bronchial and systemic reactions, *J. Allergy Clin. Immunol.*, 89, 759, 1992.
40. Mäkinen-Kiljunen, S., Alenius, H., Palosuo, T., Reunala, T., and Turjanmaa, K., Measurement of latex IgE and IgG antibodies with RAST, *Prog. Proc. Int. Latex Conf. Sensitivity to Latex in Medical Devices*, Baltimore, 1992, 39.
41. Cronin, E., Clinical patterns of hand eczema in women, *Contact Derm.*, 13, 153, 1985.
42. Turjanmaa, K. and Reunala, T., Healthcare workers. A risk group for latex allergy, *Prog. Proc. Int. Latex Conf. Sensitivity to Latex in Medical Devices*, Baltimore, 1992, 7.
43. von Baur, X., Kerz, G., and Schueuermann, M., Berufsbedingte Allergie gegen Latex, *Arbeitsmed. Sozialmed. Präventivmed.*, 28, 19, 1993.
44. Maso, M. J. and Goldberg, D. J., Contact dermatoses from disposable glove use. A review, *J. Am. Acad. Dermatol.*, 23, 733, 1990.
45. Fuchs, T. and Wahl, R., Immediate reactions to rubber products, *Allergy Proc.*, 13, 61, 1992.
46. Belsito, D. V., Contact urticaria caused by rubber, Analysis of seven cases, *Dermatol. Clin.*, 8, 61, 1990.
47. Dooms-Goossens, A., Contact urticaria caused by rubber gloves (letter), *J. Am. Acad. Dermatol.*, 18, 1360, 1988.
48. Turjanmaa, K., Incidence of immediate allergy to latex gloves in hospital personnel, *Contact Derm.*, 17, 270, 1987.
49. Beaudouin, E., Pupil, P., Jacson, F., Laxenaire, M. C., and Moneret-Vautrin, D. A., Allergie professionelle au latex, Enquete prospective sur 907 sujets du milieu hospitalier, *Rev. Fr. Allergol.*, 30, 157, 1990.
50. Cormio, L., Turjanmaa, K., Talja, M. T., Andersson, L. C., and Ruutu, M., Toxicity and immediate allegenicity of latex gloves, *Clin. Exp. Allergy*, 23, 618, 1993.
51. Kanerva, L., Jolanki, R., and Toikkanen, J., Työperäiset allergiat vähenivät hieman vuonna 1991. (Official statistics on occupational diseases and occupational injuries in Finland, Institute of Occupational Health, Helsinki), *Suomen Lääkärilehti*, 93, 40, 1992.
52. Rystedt, I., Factors influencing the occurrence of hand eczema in adults with a history of atopic dermatitis in childhood, *Contact Derm.*, 12, 185, 1985.
53. Slater, J. E., Mostello, L. A., and Shaer, C., Rubber-specific IgE in children with spina bifida, *J. Urol.*, 146, 578, 1991.

54. Lezaun, A., Marcos, C., Martin, J. A., Quirce, S., and Gomez, M. L. D., Contact dermatitis from natural latex, *Contact Derm.,* 27, 334, 1992.
55. Moneret-Vautrin, D. A., Laxenaire, M. C., and Bavoux, F., Allergic shock to latex and ethylene oxide during surgery for spina bifida, *Anesthesiology,* 73, 556, 1990.
56. Assalve, D., Cicioni, C., Perno, P., and Lisi, P., Contact urticaria and anaphylactoid reaction from cornstarch surgical glove powder, *Contact Derm.,* 19, 61, 1988.
57. Fisher, A. A., Contact urticaria and anaphylactoid reaction due to corn starch surgical glove powder, *Contact Derm.,* 16, 224, 1987.
58. van der Meeren, H. L. M. and van Erp, P. E. J., Life-threatening contact urticaria from glove powder, *Contact Derm.,* 14, 190, 1986.
59. Seggev, J. S., Mawhinney, T. P., Yunginger, J. W., and Braun, S. R., Anaphylaxis due to cornstarch surgical glove powder, *Ann. Allergy,* 65, 152, 1990.
60. Turjanmaa, K., Reunala, T., Alenius, H., Brummer-Korvenkontio, H., and Palosuo, T., Allergens in latex surgical gloves and glove powder (letter), *Lancet,* 336, 1588, 1990.
61. Slater, J. E. and Chhabra, S. K., Latex antigens, *J. Allergy Clin. Immunol.,* 89, 673, 1992.
62. Chambeyron, C., Dry, J., Leynadier, F., Pecquet, C., and Tran Xuan Thao, Study of the allergenic fractions of latex, *Allergy,* 47, 92, 1992.
63. Alenius, H., Turjanmaa, K., Palosuo, T., Mäkinen-Kiljunen, S., and Reunala, T., Surgical latex glove allergy: characterization of rubber protein allergens by immunoblotting, *Int. Arch. Allergy Appl. Immunol.,* 96, 376, 1991.
64. Morales, C., Basomba, A., Carreira, J., and Sastre, A., Anaphylaxis produced by rubber glove contact. Case reports and immunological identification of the antigens involved, *Clin. Exp. Allergy,* 19, 425, 1989.
65. Warpinski, J. R., Folgert, J., Cohen, M., and Bush, R., Allergic reaction to latex. A risk factor for unsuspected anaphylaxis, *Allergy Proc.,* 12, 95, 1991.
66. Mäkinen-Kiljunen, S., Turjanmaa, K., Palosuo, T., and Reunala, T., Characterization of latex antigens and allergens in surgical gloves and natural rubber by immunoelectrophoretic methods, *J. Allergy Clin. Immunol.,* 90, 230, 1992.
67. Ross, B. D., McCullough, J., and Ownby, D. R., Partial cross-reactivity between latex and banana allergens, *J. Allergy Clin. Immunol.,* 90, 409, 1992.
68. M'Raihi, L., Charpin, D., Pons, A., Bongrand, P., and Vervloet, D., Cross-reactivity between latex and banana, *J. Allergy Clin. Immunol.,* 87, 129, 1991.
69. Lavaud, F., Cossart, C., Reiter, V., Bernard, J., Deltour, G., and Holmquist, I., Latex allergy in patient with allergy to fruit, *Lancet,* 339, 492, 1992.
70. Wrangsjö, K., Montelius, J., and Eriksson, M., Teats and pacifiers — an allergy risk for infants?, *Contact Derm.,* 27, 192, 1992.
71. Mäkinen-Kiljunen, S., Sorva, R., and Juntunen-Backman, K., Latex dummies as allergens, *Lancet,* 339, 1608, 1992.

18

Other Reactions from Gloves

James S. Taylor

TABLE OF CONTENTS

I. Introduction 255
II. Irritant Reactions 255
III. Other Urticaria 258
IV. Chemical Leukoderma 259
V. Endotoxins 261
VI. Ethylene Oxide 262
References 264

I. INTRODUCTION

Classic allergic reactions to gloves occur as immunologic contact urticaria (ICU) to latex proteins and as allergic dermatitis to the major rubber accelerators and antioxidants. Among the easiest adverse reactions to document,[1] they are a particular problem for people whose occupations demand that they wear gloves. Less commonly reported glove reactions also occur. This chapter will discuss those other reactions, both irritant and allergic. The true incidence of irritant reactions to gloves is unknown, but they may be the most frequent of reported reactions. Most of the allergic reactions discussed are probably infrequent or rare. Glove reactions are classified in Table 1.

II. IRRITANT REACTIONS

There are no published studies of glove irritation, and the subject is only briefly mentioned in the recent literature. Heese et al.[3] state that irritant reactions occur especially in atopic patients and may be mechanically provoked by glove powder crystals; there is no reference for the latter assertion. Most familiar is the use of glove powders as donning lubricants. Corn starch is commonly used today and has replaced club moss *(Lycopodium)* powder and talc.[4,5] Lactose

TABLE 1 Classification of Glove Reactions

Irritation from occlusion, friction and maceration; worse with dermographism
Allergy to glove material
  Contact dermatitis
  Contact urticaria, angioedema and anaphylaxis
Aggravation of pre-existing dermatoses by gloves
Penetration of chemicals through gloves — epoxy, acrylics, permanent wave chemicals, etc.
Other:
  Endotoxin reaction[28]
  Ethylene oxide reactions[29-35]
  Chemical leukoderma[18-27]

Adapted from Estlander, T. and Jolanski, R., *Dermatol. Clin.*, 6, 105, 1988.

also is occasionally used, and on some gloves, silicone is used. In addition, glove powders are employed in latex glove dipping to facilitate glove removal from porcelain molds. According to Dr. D. Hogan (personal communication, May 1991), Tolbert and Brown's[4] study of glove powder showed that because powder is applied before gloves are fully cured, particles become embedded in the soft rubber. Thus, once powdered, a glove can never be made powder free.

Gloves can produce occlusion and maceration, especially following prolonged use, which are major factors in glove irritation.[3,6,7] Compounding these may be the practice of double gloving for infectious cases, friction from gloves rubbing against skin, and frequent hand washing with surgical scrubs, and brushes. Cutaneous exposure to soaps, detergents, and other cleaning agents are also major factors in the high prevalence of irritant dermatitis in occupations such as hospital cleaning.[3] Rietschel has emphasized that in certain industrial situations gloves may serve as wicks, absorbing chemicals and then aggravating or producing dermatitis.[8] Rarely is bacterial endotoxin release by gamma irradiated gloves and ethylene oxide sterilized gloves, an irritant factor[3] (see separate discussion of these two entities). Penetration of chemicals through gloves and aggravation of existing skin disease are other contributing causes.[2] Mathias[9] believes that gloves should not be worn over inflamed skin unless worn for short periods and topical corticosteroids are first applied. Irritation may also be caused by increased glove use and abuse from over use in unnecessary tasks. For example, one study found glove use was appropriate at rates of only 59% on hospital wards vs. 90% in the laboratory. Only 52% of nurses washed their hands upon doffing the gloves.[10] Variations in the thickness and tightness of examination, surgical, utility and industrial gloves may be factors in producing irritation. The type of glove material may also be important (Table 2). The irritant potential, if any, of the newer cut-resistant glove liners made of stainless steel, Kevlar® (DuPont), Lycra® (DuPont), Spectra® (Allied Signal) and other materials should be studied.[11]

Some of the published literature on irritation is relevant to glove reactions. Lammintausta and Maibach[12] classify irritation from all causes into eight clinical types: (1) acute irritant dermatitis (primary irritation), (2) irritant

TABLE 2   Glove Materials

Rubber polymers — (cause of most glove reactions), natural, butyl, nitrile, Neoprene® and Viton®
Plastic polymers — polyvinylchloride, polyethylene and polyvinylalcohol
Multilayered laminate — ethylene vinyl alcohol co-polymer is laminated with polyethylene on both sides (4H™-gloves)
Leather — chrome or vegetable tanned
Textiles — natural or synthetic; woven from fabrics, knit or terry cloth. May be coated with rubber or plastic.
Special material — wire cloth — stainless steel or nickel; e.g., used for autopsy prosectors

Adapted from Estlander, T. and Jolanski, R., *Dermatol. Clin.,* 6, 105, 1988.

reactions, (3) delayed acute irritant dermatitis, (4) cumulative irritant contact dermatitis, (5) traumatic irritant dermatitis, (6) pustular and acneiform dermatitis, (7) nonerythematous irritation, and (8) subjective irritation. Table 3 lists the types and causes of irritation as well as examples of possible glove-associated dermatitis.

Clinical features that may suggest a specific direct chemical, physical, or mechanical cause of irritant dermatitis include (1) ulceration, (2) folliculitis and acne, (3) miliaria, (4) pigmentary changes, (5) alopecia, (6) urticaria, and (7) granulomas.[12] Of these, gloves may directly cause or have been reported to cause ulceration from severe ethylene oxide (ETO) burns (see section on ETO); folliculitis, miliaria, or both from friction or occlusion; hypopigmentation from glove antioxidants, usually an immunologic reaction, (see section on chemical leukoderma); ICU from natural rubber latex proteins; and possibly nonimmunologic contact urticaria from glove powder.[1-3] In other cases gloves may indirectly aggravate ulceration caused by chemicals or directly prevent ulceration, pigmentary changes, urticaria, and granulomas by acting as a barrier between chemicals and the skin.

Noncutaneous granulomas from talc have been reported in wound scar and intra-abdominal adhesions. Talc has been replaced by starch, but granulomas from starch also have been reported in the peritoneum, pleura, pericardium, synovium, cranium, and eye.[5] Starch-free gloves are now available, such as Bio Gel® D (Reagent Hospital Products), Pristine® (World Medical Supply), Puritee® (Orox), and Neolon® (Becton Dickinson) (D. Hogan, personal communication, May 1991).

The diagnosis of irritant dermatitis is generally based on the exclusion of allergic contact dermatitis and immunologic contact urticaria. Rietschel's[13] criteria for irritant dermatitis include: (1) a rapid onset of symptoms (minutes to hours); (2) discomfort, especially in the early stages (especially stinging and burning); and less so (3) onset of dermatitis within 2 weeks of exposure; and (4) the identification of other persons similarly affected.[14]

In Maibach's view (H. I. Maibach, personal communication, February 1993), irritant reactions specific to gloves are suggested by three observations. First patch testing with pieces of some gloves results in macular erythema at

TABLE 3 Types and Examples of Irritation and Associated Glove Dermatitis

Type of irritation	Examples of direct chemical, physical, or mechanical causes	Examples of possible associated skin reactions to gloves
Acute irritant dermatitis	Strong acid or alkali burn	Excess ETO in gloves
Irritant reactions	Wet work; perm chemicals	Chemicals penetrating gloves
Delayed acute irritant reaction	ETO and HF burns; hexanediol diacrylate	Excess ETO in gloves; HF penetrating gloves
Cumulative irritant contact dermatitis	Wet work	Sweating and maceration from gloves; chemical gloves
Traumatic irritant dermatitis	Burn; laceration	Aggravation by gloves?
Pustular and acneiform dermatitis	Tar, oil, grease	Acne mechanica from glove or gauntlet friction
Nonerythematous irritation	Cosmetics; textiles	Glove occlusion as a direct cause?
Subjective irritation	Lactic acid	Enhancement by gloves?

Note: ETO = ethylene oxide; HF = hydrofluoric acid.
Modified from Lammintausta, K. and Maibach, H. I., *Occupational Skin Diseases,* Adams, R. M., Ed., W. B. Saunders, Philadelphia, 1990, chap. 1.

48 h that largely resolves at 96 h. A similar patch test response occurs in 21 day cumulative assay studies. Also some patients with neither allergic contact dermatitis nor immunologic contact urticaria can tolerate only a latex glove with a liner. Silk, nylon, and cotton liners are available. Silk (Sensi Touch®) liners are claimed by the manufacturer to preserve tactile sensation while absorbing sweat and reducing friction.[15]

Glove irritation may be a significant cause of glove reactions but needs further detailed study with *in vitro, in vivo,* and predictive cumulative irritation studies. All glove components and materials should be investigated.

III. OTHER URTICARIA

According to Heese et al.[3] nonimmunologic contact urticaria (NICU) from gloves has not been reported, but sorbic acid, a known NICU agent is sometimes added to glove powder and should be considered in evaluating patients with glove-induced contact urticaria.

Pressure urticaria and cholinergic urticaria associated with glove use should be diagnosed by the patient's history and cutaneous examination.[3] Pressure urticaria, a form of physical urticaria, is either localized at the stimulus site (contact type) or associated with distant reactions, such as hypotension, tachycardia, tickle, or pruritus of the nose or finger tips (reflex type).[16] Atopy is frequently absent and lesions do not occur at night. Pressure urticaria arises after a single stimulus, and the skin reaction is not a typical wheal but rather a reddened, deeper, sometimes painful swelling of the angioedema type. Urticarial dermographism may be associated with pressure urticaria. Braun Falco et al.[16] describe both immediate (3 to 30 min) and delayed (2 to 6 h) urticarial dermographism and pressure urticaria. Pressure urticaria may be immunologic.

Cholinergic (sweat or exertion) urticaria follows an increase in body temperature. Tiny, confluent, pruritic wheals on a patchy erythema develop; lesions usually occur on the upper body after physical exertion or sweating and urticarial dermographism may be present.[16]

Distinguishing irritant reactions from NICU may be difficult. Lahti[17a] points out that strong irritants, such as hydrochloric acid, formaldehyde, and phenol, can cause immediate wheaving followed by erythema, and either scaling or crusting, which lasts 20 h or longer. Nicotinic acid esters produce only contact urticaria; sodium lauryl sulfate is a pure irritant while phenol and dimethyl sulfoxide have both features. Diagnostic tests for NICU and ICU include the rub, use, open application, patch, and chamber tests, usually performed in that order. Prick, RAST, and passive transfer tests are negative for NICU.

In 1984, Anderson and Maibach[17b] described multiple application, delayed-onset contact urticaria to formalin as possibly related to certain textile reactions. The reaction appeared on normal skin after repeated open applications but only after a single application on diseased skin. Although they described this reaction as possibly a nonclinically relevant epiphenomenon, it could be relevant to unexplained glove reactions.

IV. CHEMICAL LEUKODERMA

The first cases of chemical leukoderma occurred in tannery workers wearing heavy rubber gloves and were reported in 1939 and 1940 by Schwartz, Oliver, and Warren of the U.S. Public Health Service.[18,19] These reports are classics in the annals of occupational dermatology. Depigmentation of the skin was caused by the rubber antioxidant "Agerite Alba", also known as monobenzylether of hydroquinone (MBEH). In one tannery 25 (52%) of 48 workers who wore heavy acid-cured rubber gloves with MBEH present at 0.5% were affected. The workers wore the rubber gloves for many months before the leukoderma developed. According to the 1940 report, "itching and in some case mild dermatitis" preceded the appearance of the depigmentation without "great discomfort". However in a 1947 review, Schwartz[20] stated, "There was [sic] no inflammatory symptoms at any time." This discrepancy may be explained by the 1940 report[19] that no signs of acute or chronic dermatitis were present in the areas of leukoderma at the time of examination. In some workers the leukoderma occurred on the hands and half way up the forearms with a uniform sharp cut-off line corresponding to areas covered by the gloves. In others there was a patchy, guttate, confetti-like pigment loss. The face and trunk were also involved in some workers, probably from direct contact with the gloves. Hair in the leukodermic areas remained pigmented. The depigmentation was most marked in African Americans, less intense in Hispanics, and noticeable in Caucasians only in the summer.

Patch testing with various chemicals in the gloves confirmed MBEH as the cause.[19] Eight different patch tests with unspecified concentrations of the glove ingredients were occluded for 7 days on 10 workers. When one of the MBEH patch test sites (site #3) was read at 14 days on each of the 10 workers, all showed a positive eczematous reaction and 6 of the 10 showed leukoderma. After 6 months 3 more workers had developed leukoderma at the patch test sites. There was no direct "correlation between intensity of skin reaction to the MBEH patch tests and the subsequent development of leukoderma."[19] Other MBEH patch tests occluded for 72 h caused leukoderma to develop 3 to 5 months after the patches were removed. Hydroquinone was present as an impurity at less than 1% and was not a factor in the leukoderma.

Several months after workers stopped using the gloves, partial repigmentation occurred. Repigmentation was perifollicular, spreading peripherally; was more rapid in some than in others; and was "practically completed in all the cases three years later. The general health of the workers was not affected."[19] Further investigation by the U.S. Public Health service showed that a considerable percentage of workers wearing the same make of gloves in the tanneries and in other industries was also affected. According to Ortonne, Mosher, and Fitzpatrick, in 1939 McNally[21] reported depigmentation in 34 employees in a tannery. Leukoderma also occurred at remote sites and was attributed to accidental direct contact with the gloves. In 1959 Botvinick[23] reported dermatitis in secondary leukoderma from a fabric-lined household glove. MBEH was identified in one of the gloves, and the dermatitis and leukoderma were reproduced by patch testing.

In 1985 I saw a patient with chemical leukoderma of the hands from a glove manufactured in Germany and distributed in the U.S. The company stated that MBEH was not present in the glove, and the cause of the leukoderma remains unknown. In 1992, Bajaj, Gupta, and Chatterjee[25] identified a patient with depigmentation at the site of a hearing aid. High-pressure liquid chromatography (HPLC) revealed that the hearing aid contained MBEH.

Leukoderma caused by MBEH has been reported from a number of other rubber devices including tape, diaphragms, condoms, finger cots, clothing, aprons, dolls, and shoes.[23,24] It has also been identified in synthetic Neoprene® rubber.[24] The U.S. rubber industry has not used MBEH as an antioxidant for many years.[1]

Related chemicals producing pigment loss include hydroquinone,[23] monomethylether of hydroquinone (p-methoxyphenol or p-hydroxyanisole) and monoethylether of hydroquinone (p-ethoxyphenol). Hydroquinone rarely produces complete depigmentation, does not produce pigment loss at distant sites as does MBEH, and is a weaker allergen than MBEH. A number of other depigmenting chemicals have now been identified, especially the alkylphenol *p*-tertiarybutylphenol (PTBP). When present in excess in Neoprene® rubber glue-containing *p*-tertiarybutylphenol formaldehyde resins, PTBP has been associated with depigmentation in industrial workers and in consumers when present in wristwatch adhesive and plastic shoes.[1,23] To my knowledge, PTBP has not been found in gloves.

Chemical leukoderma may appear identical to idiopathic vitiligo and may have a similar antitomic distribution.[26] The "mode of spread may be helpful — a history of gradual coalescence of small discrete macules, rather than development of large macules with perifollicular sparing suggests chemical leukoderma".[24] Scalp hair is rarely involved and eye color does not change. The incubation period for exposure ranges from 2 weeks to approximately 6 months.[26] Depigmentation is not always preceded by inflammation of the affected skin but is frequently associated with allergic contact dermatitis to the same chemical responsible for the pigment loss, although the latter is not a prerequisite.[21] The absence of preceding inflammation may be explained by the resistance of some idiopathic vitiligo patients to developing a contact allergy in depigmented skin sites.[27] Wood's light examination of the skin in a dark room may identify areas of leukoderma not obvious on routine visual inspection of the skin.

Diagnosis of chemical leukoderma is more easily made when a number of cases are clustered in a factory, and there is a history of worker exposure to a known depigmenting agent; when pigment loss follows contact dermatitis; or when the person affected is an adult without childhood or family history of vitiligo.[26] Patch tests with putative depigmenters may result in leukoderma at the sites of positive or negative tests for up to 6 weeks or more after application. Satellite depigmentation may occur, and patch testing should be done cautiously (I prefer to patch test on the buttocks) with unknown chemicals. Application of the putative chemicals to black guinea pigs[26] or identification of unknown depigmenters by HPLC[25] are alternative test methods.

V. ENDOTOXINS[28]

Endotoxin reactions are "an ever-present danger in the preparation of biologicals for intravenous and intramuscular use."[28] The problems associated with endotoxin reactions are usually systemic and are characterized by fever, chills, and hypotension. Local reactions may also occur. Generalized and localized Schwartzman reactions "are older terms used in reference to these reactions." The pathogenesis of fever in endotoxin reactions "is due to the release of endogenous pyrogens from activated granulocytes rather than a direct response to bacterial endotoxin. The exogenous stimuli can be bacteremia, viremia or endotoxemia."

Shmunes and Darby[28] reported the only known case of contact dermatitis apparently due to endotoxin present in irradiated latex gloves worn by a hospital phlebotomist. The patient developed "vivid erythema along the sides and back of the finger shafts with vesiculation and relative sparing of the palms."[28] The source of the irritation, which occurred clinically and on usage tests 12 to 24 h after wearing gloves (Micro Touch® by Arbrook, Inc.) for 10 to 20 min, was linked to bacterial endotoxin in sterile latex gloves. The reaction repeatedly recurred within the same time frame. On one occasion after handling a rubber tourniquet enclosed in a sterile i.v. catheter packed for several minutes, isolated patches and linear streaks developed under the patient's chin, neck, cheek, and eyelid. A rubbing test using a tourniquet from a freshly opened catheter pack produced a positive response. Patch tests done with the standard screening tray were negative. The gloves, however, were not patch tested. There was no immediate itching within the first hours suggestive of contact urticaria. The lag time of at least 12 h between testing and the development of the erythema and vesicles made delayed contact urticaria seem unlikely. RAST or prick testing was not performed. Glove samples from the same lot were analyzed and all were found to contain endotoxin levels higher than permitted. Sterilization of the gloves by gamma irradiation was reported to actually increase endotoxin levels when the bacterial count was elevated. Water soluble endotoxin was absorbed onto powder inside the gloves, and sweating under the gloves may have enhanced entry into the skin. No systemic symptoms which may occur in cases of endotoxin reactions from contaminated biologicals intended for intramuscular or i.v. use were reported. This report deserves further study. Shmunes and Darby[29] postulate that dihydrosis from cutting fluids could also be based on bacterial endotoxins.

VI. ETHYLENE OXIDE

Ethylene oxide (ETO) was first used as a sterilizing agent for medical supplies in 1962. The ability of ETO to sterilize depends on its alkylating properties and its irreversible bactericidal effect on cell metabolism. Over the years it has been used to sterilize a number of reusable medical supplies susceptible to heat, such as those made of plastic and rubber, in which ETO is soluble and retained in large amounts after sterilization. Hazards associated with inadequate aeration of ETO sterilized devices and equipment include cutaneous burns (e.g., from rubber gloves), tracheal inflammation (e.g., endotracheal tubes), hemolysis (e.g., from plastic tubing) and anaphylaxis (e.g., plastic and rubber tubing used for hemodialysis).[29] Adequate aeration of the devices is therefore imperative to ensure that all ETO residues are eliminated.[29-31]

There have been reports of ETO irritation and burns from industrial[32] and medical gloves.[33] Royce and Moore[33] report ETO irritation in microbiology workers. A hermetically sealed glove box, devised for performing aseptic operations in a microbiology lab, was sterilized with ETO vapor. When the box's long rubber gloves were used without adequate aeration, all of the operators developed dermatitis. The problem was solved by extending the whole of the glove and gauntlet outside of the box and allowing the rubber to "air off" for 1.5 h before work started.[32] Rendell-Baker[34] cites instances of burns on surgeons' hands from ETO residues in gloves. In another case, he cites hand irritation from ethylene chlorhydrin and ethylene glycol residues in ETO-sterilized gloves.[35] Other studies have shown these two compounds to be much less irritating than ETO.[29] In 1988, Fisher[33] reported ETO burns on the hands of a hospital worker wearing heavy-duty rubber gloves.

In addition, ETO burns have been reported from sterilization of anesthesia masks, hospital linens, endotracheal tubes and other medical devices. Commercially sterilized nitrofurozone gauze dressings have also been cited as a cause of first and second-degree burns. In addition, a number of industrial workers exposed to ETO have suffered burns, irritant contact dermatitis, and allergic contact dermatitis.[29]

ETO is used much less than in the past and most surgical gloves are sterilized by gamma irradiation. If ETO is used, adequate aeration is imperative and is probably best handled by mechanical aeration. Special handling of polyvinylchloride plastic, rubber, and previously gamma irradiated equipment is indicated. Not only is ETO a potent irritant, but it also can cause delayed hypersensitivity (allergic contact dermatitis) and immediate hypersensitivity (anaphylaxis).[29,31,32]

In 1986, I saw an operating room nurse with an allergic contact dermatitis of the hands that occurred only after she wore ETO-sterilized gloves. Her patch test reactions to the standard tray, rubber chemicals, a steam-sterilized glove provided by the same manufacturer, and to the gamma irradiated gloves were all negative. Her patch test reaction was positive to ETO-sterilized gloves only. These patch tests were confirmed several months later with pieces from the same ETO-sterilized glove previously tested. Other batches of the same ETO-sterilized gloves and other brands of ETO-sterilized gloves produced the same clinical and patch test reactions. Her only alternative was to wear gamma irradiated gloves. Highly reactive, ETO can combine with a number of other chemicals, such as rubber accelerators and iodine. The reaction in this nurse may have been caused by ETO itself or by an ETO rubber-chemical reaction product.[1] Downy showed that the mercaptobenzothiazole vulcanization accelerators in rubber react rapidly with ETO to produce (hydroxyethyl-mercapto)-thioazole, despite the fact that residual ETO concentration in the rubber tubing have been reported to dissipate after 5 h of aeration. Thus, small amounts of free ETO may remain in the rubber matrix and can be a contributing factor in allergic contact dermatitis.[36]

Patch testing for thresholds of ETO irritation and allergy has been carried out with ETO impregnated materials such as gauze, rubber, PVC, and petrolatum.[31] Fisher[31] reviews these studies, one of which revealed an irritation threshold of 1000 ppm ETO in thick PVC or petrolatum and atypical mild delayed allergy at 100 ppm in PVC.[37]

REFERENCES

1. Taylor, J. S., Rubber, in *Contact Dermatitis,* Fisher, A. A., Ed., Lea & Febiger, Philadelphia, 1986, chap. 36.
2. Estlander, T. and Jolanki, R., How to protect the hands, *Dermatol. Clin.,* 6, 105, 1988.
3. Heese, A., Hintzenstern, J. V., Peters, K. P., Koch, H. U., and Hornstein, O. P., Allergic and irritant reactions to rubber gloves in medical health services, *J. Am. Acad. Derm.,* 25, 831, 1991.
4. Tolbert, T. W. and Brown, J. L., Surface powders on surgical gloves, *Arch. Surg.,* 115, 729, 1980.
5. Beck, W. C., Issues related to surgical gloves, *Biomed. Instr. Technol.,* 26, 225, 1992.
6. Agrup, G., Hand eczema, *Acta Dermatol. Venereol.,* 49 (Suppl. 61). 1, 1969.
7. Fay, M. F. and Sullivan, R. W., Changing requirements for glove selection and hand protection, *Biomed. Instr. Technol.,* 26, 227, 1992.
8. Rietschel, R. L., Irritant Contact Dermatitis Lecture, *Symp. Occupational Skin Disease,* American Academy of Dermatology, December, 1988.
9. Mathias, C. G. T., Treatment of Occupational Contact Dermatitis, Syllabus, *Occupational Skin Disease Sem.,* American College of Occupational and Environmental Medicine, April 1993.
10. DeGrot-Kosolcharoen, J., Pandemonium over gloves: use and abuse, *Am. J. Infect. Contr.,* 19, 225, 1991.
11. Witmeyer, A. J., Trends in glove product and process development, *Biomed. Instr. Technol.,* 26, 235, 1992.
12. Lammintausta, K. and Maibach, H. I., Contact Dermatitis Due to Irritation, in *Occupational Skin Diseases,* 2nd ed., Adams, R. M., Ed., W. B. Saunders, Philadelphia, 1990, chap. 1.
13. Rietschel, R. L., Patch testing and occupational hand dermatitis, *Dermatol. Clin.,* 6, 43, 1988.
14. Maso, M. J. and Goldberg, D. J., Contact dermatoses from disposable glove use: a review, *J. Am. Acad. Dermatol.,* 23-733, 1990.
15. Field, E. A. and King, C. M., Skin problems associated with routine wearing of protective gloves in dental practice, *Br. Dent. J.,* 168, 281, 1990.
16. Braun-Falco, O., Plewig, G., Wolff, H. H., and Winkelmann, R. K., *Dermatology,* Springer-Velag, Berlin, 1991, chap. 11.
17a. Lahti, A. and Maibach, H. I., Immediate contact reactions, *Immunol. Aller. Clin. N. Am.,* 9, 463, 1989.
17b. Anderson, K. E. and Maibach, H. I., Multiple application delayed onset contact urticaria, *Contact Derm.,* 10, 227, 1984.
18. Oliver, E. A., As referenced in Ortonne, J.-P., Mosher, D. B.. and Fitzpatrick, T. B., Chemical hypomelanosis, in *Vitiligo and Other Hypomelanoses of Hair and Skin,* Plenum Press, New York, chap. 5, 1993.
19. Schwartz, L., Oliver, E. A., and Arren, L. H., Occupational leukoderma, *Public Health Service Rep.,* 55, 1111, 1940.
20. Schwartz, L., Occupational pigmentary changes in the skin, *Arch. Dermatol. Syph.,* 56, 592, 1947.

21. McNally, As referenced in Ortonne, J.-P., Mosher, D. B., and Fitzpatrick, T. B., Chemical hypomelanosis, in *Vitiligo and Other Hypomelanoses of Hair and Skin,* Plenum Press, New York, chap. 5, 1993.
22. Botvinick, I., Dermatitis and secondary leukoderma due to fabric-lined rubber gloves, *Arch. Dermatol. Syph.,* 53, 334, 1951.
23. Fisher, A. A., Vitiligo due to contactants, *Cutis,* 17, 431, 1976.
24. Ortonne, J.-P., Mosher, D. B., and Fitzpatrick, T. B., Chemical hypomelanosis, in *Vitiligo and Other Hypomelanoses of Hair and Skin,* Plenum Press, New York, chap. 5, 1993.
25. Bajaj, A. K., Gupta, S. C., and Chatterjee, A. K., Hearing aid depigmentation, *Contact Derm.,* 27, 126, 1992.
26. Gellin, G. A. and Maibach, H. I., Chemically induced depigmentation, in *Models in Dermatology,* Maibach, H. and Low, N., Eds., Karger, Basel, 1985.
27. Uehara, M., Miyauchi, H., and Tanaka, S., Diminished contact sensitivity response in vitiliginous skin, *Arch. Dermatol.,* 120, 195, 1984.
28. Shmunes, E. and Darby, T., Contact dermatitis due to endotoxin in irradiated latex gloves, *Contact Derm.,* 10, 240, 1984.
29. Taylor, J. S., Dermatologic hazards from ethylene oxide, *Cutis,* 19, 189, 1977.
30. Glaser, Z. R., Use of Ethylene Oxide as a Sterilant in Medical Facilities, DHEW (NIOSH) Publ. No. 77-200, NIOSH, CDC, Cincinnati, Ohio, 1977.
31. Fisher, A. A., Ethylene oxide dermatitis, *Cutis,* 34, 20, 1984.
32. Royce, A. and Moore, W. K. S., Occupational dermatitis caused by ethylene oxide, *Br. J. Ind. Med.,* 12, 169, 1955.
33. Fisher, A. A., Burns of the hands due to ethylene oxide used to sterilize gloves, *Cutis,* 42, 267, 1988.
34. Rendell-Baker, L., Roberts, R. B., and Watson, B. M., Problems in sterilization of medical equipment, *Hosp. Bureau Res. News,* 16, 1, 1969.
35. Smith, E. A., As cited by Rendell-Baker, L. in Problems in sterilization of medical equipment, *Hosp. Bureau Res. News,* 16, 1, 1969.
36. Downey, P. M., As referenced in Glasser, Z. R. (see Reference 30).
37. Shupack, J. L., Anderson, S. R., and Romano, S. J., Human skin reactions to ethylene oxide, *J. Lab. Clin. Med.,* 98, 723, 1981.

PART VI
APPLICATION OF TEST DATA

19

The Selection and Use of Gloves Against Chemicals

Paul Leinster

TABLE OF CONTENTS

I. Introduction 270
II. Factors in Glove Selection 270
III. Performance Characteristics 272
IV. Limitations in Test Procedures 272
V. Selection Procedure 273
    A. Chemical Classification 273
    B. Definition of Activity 273
    C. Glove Selection 273
    D. Glove Reuse Criteria 274
VI. Use of Gloves in the Workplace 276
    A. Glove Specification 276
    B. Assessment of Glove Suitability 276
    C. Degree of Protection Provided 276
    D. Comfort 277
    E. Choice 277
    F. Dexterity 277
    G. Period of Use 278
    H. Issue 278
    I. Supervision, Monitoring, and Review 278
    J. Decontamination and Cleaning 278
    K. Storage 279
    L. Training 279
    M. General Considerations 279
VII. Conclusions 280
References 280

I. INTRODUCTION

Gloves are often used to prevent skin contact with chemicals capable of causing local or systemic effects. It is frequently stated in health and safety data sheets that impervious or chemically resistant gloves should be worn when handling the chemicals. The ideal gloves should provide sensitive touch, resist all chemicals, keep the hands clean and at the correct temperature, and never require replacement. However, commonly available glove materials provide only limited protection against many chemicals and therefore it is necessary to select the most appropriate glove for a particular application and to provide guidance on how long it can be worn and whether it can be reused.

Within the U.K. the Management of Health and Safety at Work (MHS) Regulations 1992[1] and the Personal Protective Equipment at Work (PPE) Regulations 1992[2] came into force on January 1, 1993, implementing European Directives. The MHS Regulations require employers to identify and assess the risks to health and safety present in the workplace. The purpose of the assessment is to ensure that the most appropriate means of reducing those risks to an acceptable level are identified. Within the accompanying Approved Code of Practice to the PPE Regulations the principle of the use of personal protective equipment as a "last resort" is discussed. The code states what health and safety professionals have long advocated: that controls and safe systems of work should always be considered first. However, there are situations where control at source is not reasonably practicable or control measures have not yet been incorporated, and under these conditions personal protection is necessary to reduce to an acceptable level the exposure of the personnel involved.

Even in situations where gloves are being used it is important to stress to wearers that systems of work which minimize the potential for contact with chemicals should be adopted. Care should always be taken not to contaminate the skin, clothing, equipment, or the working environment and any spillages should be cleaned up immediately when they occur.

II. FACTORS IN GLOVE SELECTION

Personal protection is selected initially to reduce exposure to the particular hazard against which it is being used to an acceptable level. To assess its adequacy and to ensure that, in theory at least, it is capable of providing an adequate safeguard, it is necessary to obtain the following information.

Nature of the hazard — Qualitative and quantitative information is needed about the particular hazard. For example, gloves cannot be chosen for work with acid baths unless the type of acid and its concentration are known.

Performance data — Data are required regarding the ability of specific types of gloves to reduce the particular hazard. Often these data will have to be obtained from the manufacturers. Usually the data will be the result of tests performed under standardized conditions conforming to the relevant national or international standards.

Acceptable level of exposure to the hazard — For many hazards the acceptable level is zero, for example, protection of the skin from attack by caustic solutions.

A number of factors need to be taken into account when choosing a glove for a particular application. In the initial selection process the following are of primary importance:

1. The toxic properties of the chemical or chemicals should be considered. In particular, the ability of the chemical to cause local effects on the skin and/or to pass through the skin and cause systemic effects.
2. The work activities being undertaken must be studied and account taken of the degree of dexterity required, the duration, frequency, and degree of chemical exposure, and the physical stresses which will be applied.
3. The performance characteristics of the gloves should be assessed using standard test procedures. Characteristics to be considered include chemical, puncture, tear, and abrasion resistance.

The greater the potential hazard from skin contact with a chemical the greater the care required in the selection of appropriate gloves and in the development of suitable procedures for their use. In situations where the type and extent of the chemical contamination is not known, for example, when dealing with a spillage of unknown materials, then selection on the basis of a worst-case evaluation should be carried out.

The assessment used to determine the type of gloves to use in a particular situation should be documented. This will enable a review to be made if the situation changes in some way or if a particular glove proves to be unsatisfactory.

When selecting gloves the following points should be taken into account:

Liquid organic chemicals permeate through commercially available glove types.

The breakthrough time (BT) and permeation rate of a particular compound depends on factors such as glove material, thickness, temperature, pressure, stretch and aging.

When mixtures of solvents are used consideration should be given to synergistic effects enhancing the rate of permeation of one or more compounds.

Protective clothing manufactured from the same nominal material, e.g., PVC, nitrile, but supplied from different sources can have markedly different permeation properties. In fact, considerable differences in permeation rate and BT have been determined in gloves of the same nominal material, e.g., PVC, produced by a single manufacturer because of the use of different compounds, e.g., plasticizers, within the range of products.

As one would expect, the thicker the glove material the longer the BT for a particular chemical. Conversely, the thinner the glove, the less the protection offered. Therefore "disposable surgeon-type" gloves may provide very little protection against many substances.

The temperature of the chemical handled is critical. Breakthrough times decrease markedly with increasing temperature.

A significant proportion of gloves leak or have surface defects.

III. PERFORMANCE CHARACTERISTICS

The chemical resistance of glove materials is typically defined in terms of degradation, penetration, and permeation. Standardized test procedures are required so that data produced by different laboratories are comparable and these are discussed in earlier chapters.

Normally, the BT data are of primary importance in selection, but if two or more materials have similar BTs then relative permeation rates need to be considered.

Traditionally, gloves have been selected from the information given in "resistance charts" supplied by manufacturers. Such data are normally based on some form of degradation testing — although not to an agreed protocol — with no account taken of permeation. Chemicals can and do permeate materials, even if degradation (resistance) charts list the material as excellent, and even when there is no visible change in the material. Certain manufacturers are now providing BT and permeation rate data for some of their gloves against a wide range of chemicals.

IV. LIMITATIONS IN TEST PROCEDURES

The data generated by the various test procedures need to be interpreted with care. Test procedures have a number of limitations, they do not mimic the way exposure occurs in the workplace, and can be time-consuming and costly to perform. The information obtained merely provides a ranking order of glove types, for example, in terms of BTs for a particular chemical. However, it is possible to state with some certainty from test data which glove types should not be worn in a given situation.

The difficulty of providing appropriate advice still remains; for example, how long can specific types of gloves be worn in a particular application, and whether or not they should be reused.

A rationale has therefore been developed for the selection, testing, and use of gloves, taking into account the type of activity being carried out, the toxicity of the material handled, and the results from standard tests. This system provides general guidance and is not proposed as a strict regime to be adhered to in all situations. This procedure, which is described in the following sections, was first published in a paper by Leinster et al. in 1990.[3]

V. SELECTION PROCEDURE

A. Chemical Classification

The chemical or product is assigned to one of the following categories: A — not requiring labeling under the United Kingdom Classification, Packaging and Labelling of Dangerous Substances (CPL) Regulations 1984;[4] B — toxic, harmful, irritant; and C — very toxic, corrosive, carcinogenic, sensitizer, absorbed through skin.

These categories are based on the criteria given in the CPL Regulations and on the basis of these criteria chemicals not specifically listed within the regulations can be assigned to the appropriate category. However, any compounds known to be absorbed through the skin or a sensitizer should be assigned to category C.

The limitations in such a system are recognized, but it was considered that at least it provides a procedure based on readily available information. If there is any doubt regarding the category for a chemical or product it should be assigned to the higher one. Other, similar, chemical categorization systems commonly used in other countries could be readily adapted to fit this scheme.

B. Definition of Activity

The activity being undertaken is defined in terms of the likelihood of contact and the degree of exposure: (1) possible splashing; (2) occasional but known contact; and (3) known prolonged contact. Again if there is any doubt as to which category to use, the higher one should be selected.

C. Glove Selection

Using the activity and chemical classifications the required performance tests are determined by reference to Figure 1:

- Category A1 — gloves are not essential.
- Categories A2, B1, and C1 — no testing required.
- Categories A3 and B2 — a BT and permeation rate test of the type developed by the BOHS Technology Committee Working Party[5] or the American Society for Testing and Materials[6] is required: this test also provides some information on degradation characteristics.
- Categories B3, C2, and C3 — a BT and permeation rate test as above is required. In addition, to determine if pinholes are present, a test of the type given in BS 1651,[7] or a simple inflation procedure, should be carried out prior to the use of gloves for these categories.

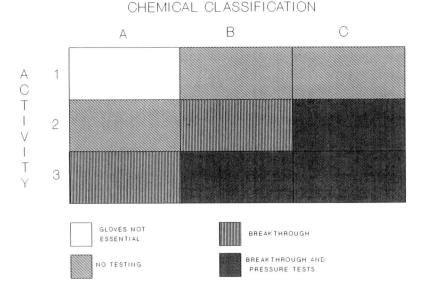

FIGURE 1 A matrix based on working activity and chemical classification gives the required glove performance tests.

In addition to the testing requirements outlined above, a simple test should be carried out to determine whether contact with the chemical causes the glove surface to become slippery. This could cause handling difficulties in activities where grip is important.

D. Glove Reuse Criteria

Once a glove type has been selected, decisions have to be made regarding how long it can be used and whether it can be reused. A matrix based on the chemical classification and activity definitions given previously is presented in Figure 2.

- Categories A1, A2, B1, and C1 — gloves can be reused as long as the surface remains intact and of a good quality. Gloves should be discarded when physical degradation, such as cracking or swelling, is apparent or they are damaged or excessively dirty. In addition, gloves in category C1 should be discarded if they are known to have been contaminated by a significant amount of chemical.
- Category A3 — gloves can be reused as long as: (1) the BT determined on a cumulative basis is not exceeded during any one period of wear; and (2) a recovery period of twice the BT is allowed between the use of a glove and its subsequent reuse. Tests should be carried out on the glove material after a number of use/recovery cycles to ensure that the BT is not significantly different from that obtained on new material. If it is, then this should be taken

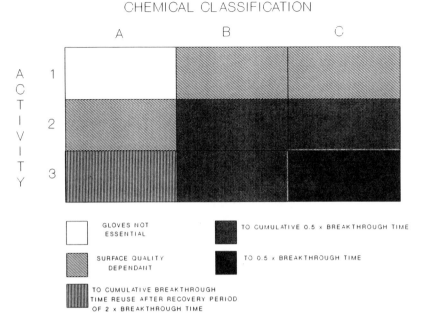

FIGURE 2 A matrix based on working activity and chemical classification gives the glove reuse criteria.

into account in defining the acceptable period of wear. In addition, the glove should be discarded if physical degradation, such as cracking, occurs.
- Categories B2, B3, and C2 — gloves can be reused for a cumulative period, up to 0.5 times the BT, after which they should be discarded.
- Category C3 — gloves can be used only within a period up to 0.5 times the BT after the first exposure, and should then be discarded.

The reason why a fraction of the BT is used for the latter categories is apparent from the way in which this value is calculated (see Figure 3). A graph of this type is used so that consistent results are obtained between laboratories.[5] It is evident that chemical has started to pass through the glove material before the quoted "breakthrough" time. Typically, however, the time at which a chemical is first detected permeating through a glove material has been found from experiment to exceed 0.5 times the BT. By inspection of the results from a specific experiment it may be possible to use a factor other than 0.5.

Other selection criteria can be used if tests are carried out or if established practice has shown them to be applicable. All the replacement times are maximum values and gloves should be replaced sooner if degradation is apparent and the physical barrier is damaged in any way, for example if cracking, hardening, abrasion, and pinholes are apparent.

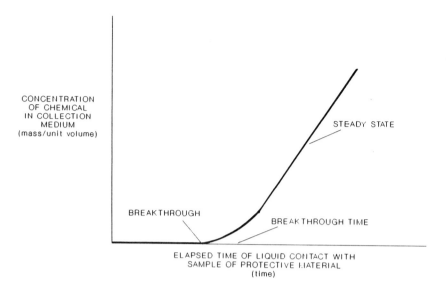

FIGURE 3 Cumulative amount of chemical permeation vs. time.

VI. USE OF GLOVES IN THE WORKPLACE

A. Glove Specification

When specifying gloves for use in a particular situation the manufacturer and specific type (type number) should be specified; it is not possible to just recommend for example "a PVC glove".

B. Assessment of Glove Suitability

When the gloves have been selected on the basis of test data and observation of the task it is then necessary to assess their performance in the work situation. Unfortunately, there are few methods available. In the case of severe failure individuals may suffer from skin problems. In other situations hands may smell of the compound when they are removed from the gloves. Visually, the inside of the gloves may appear to be contaminated or chemical analysis of special patches worn on the hand, material from the inside of the glove, or inner gloves, may indicate the presence of the chemical. In such situations investigations will be required to determine whether the chemical is permeating through the glove or if contamination is occurring in some other way. Chemical can be transferred to the insides of gloves by contaminated skin.

C. Degree of Protection Provided

The calculated degree of protection will not be achieved in practice unless the gloves are cleaned and checked regularly; those involved in the scheme receive training; the scheme is monitored; and the effect on job performance is assessed.

D. Comfort

The specified gloves must be worn at all times when the person is at risk. Therefore, the gloves must be compatible with the work which has to be done and with any other personal protection which has to be used at the same time. Comfort, or at least lack of undue discomfort, is therefore another vital factor. Different glove sizes should be made available from which the best fit for the individual user can be selected. Purchasing departments should be given instructions to obtain gloves in a range of sizes. They should not be allowed, just because they can get a better price, to purchase only one size. It is also important that they are given the detailed specification of the glove to be acquired including manufacturers name and catalog number. Gloves which theoretically provide a very high degree of protection, but are uncomfortable or do not allow the required level of dexterity and therefore are removed for part of the time, may provide less protection in practice than those which in theory are slightly less effective, but which are more comfortable and worn for a higher proportion of the time.

Most protective gloves are impervious to water vapor, and unsupported membrane types, which have no absorbent base or lining, can be unpleasant to wear for more than a few minutes as the hands become sweaty. This problem can be alleviated to some extent by the wearing of cotton inner gloves.

E. Choice

Wherever possible, users should be given a choice from a range of items, all of which must be theoretically capable of providing adequate protection. If more than one type of glove offers suitable protection for a particular operation, consideration should be given to conducting a trial to determine which is most acceptable to the work force. Personal protection will have little chance of being effective if it is not accepted by each wearer. Wearers should therefore, whenever possible, be consulted and involved in the selection of equipment.

F. Dexterity

A common reason given by people for not wearing gloves is that they do not allow sufficient dexterity and can make a job more difficult to perform. However, people get used to wearing gloves. Maintenance personnel will often wear chrome leather, rather than PVC, gloves when carrying out tasks because they consider that it is not possible to carry out mechanical manipulations when wearing impervious gloves. This is exemplified by a situation where many on a maintenance staff wore leather gloves when handling products on an oil refinery, some of which contained benzene. The gloves frequently became wetted with the chemicals. However, when carrying out similar tasks to those which they insisted could not be carried out using impervious gloves, but working on plant equipment containing sulfuric acid, all of them wore PVC gauntlets. When this was pointed out, they realized that it was all to do with perceptions of risk. If the risk was perceived to be high, as in the case with the sulfuric acid, they were willing to carry out tasks using the "less convenient" PVC gloves.

G. Period of Use

In practice it is usually more appropriate to specify glove usage in terms of wear for the duration of a particular activity or for a period related to the working day, for example half-shift, full-shift, week, rather than by use of the actual time determined by a test procedure.

H. Issue

Once gloves have been selected and guidance provided on the period of time for which they may be worn, care must be taken — and appropriate systems introduced — to ensure that the correct type of glove for a particular operation is actually issued, used, and then replaced at the required frequency. In certain situations a great amount of time and care goes into the selection of a particular glove and on specifying how it should be used. This is only part of the process, care must then be taken to ensure that the system actually works in practice. This involves instituting procedures to ensure people know why they need to wear gloves, the issue of gloves, and the training of people on all aspects of a glove use and care program.

I. Supervision, Monitoring, and Review

When an operation is introduced that requires gloves to be worn, close supervision and monitoring may be necessary initially to deal with any difficulties that occur.

The usage of gloves should be monitored visually at regular intervals. If gloves are not being worn for designated tasks then the reasons should be determined. Such checks could identify the need for a different type of glove or for additional training. The use of gloves should be kept under review and alternatives to their use found whenever possible.

Experience shows that management and supervision personnel must be committed to the use of personal protection if such a procedure is to be effective and they must lead by example.

J. Decontamination and Cleaning

Care must be taken when removing gloves to ensure that skin contamination does not occur. Single-use gloves can be removed by peeling the glove off the hand inside out. Gloves should be decontaminated before they are removed. The decontamination procedure used will depend upon the chemical contaminants. Correct decontamination and cleaning procedures should be developed as part of a comprehensive glove use program.

When gloves have been removed the hands should be washed with soap and water and then dried. Hand cream should then be applied to prevent the skin becoming dry from the loss of natural oils. Cuts and abrasions should be covered with waterproof plasters and the dressing should be changed for a porous one after work.

K. Storage

Gloves should not be left, when not being used, in places where they are likely to become contaminated, especially on the inside of the glove.

L. Training

Within the training provided it is important to deal with the misconceptions that wearers can have. Many wearers will consider that impervious gloves provide absolute protection and therefore that prolonged contact and total immersion are unimportant. Wearers should be made aware of why they must wear gloves. It is only in recent years that many health and safety professionals have become aware of the need for care in the selection and use of gloves. This awareness now has to be communicated to those at risk, the wearers of the gloves.

Experience has shown that training is required in all aspects of glove selection, issue, use, in the removal of gloves in such a way as to avoid contamination, how to check for defects, decontamination, storage, and replacement. Written operating procedures should also be prepared. In addition, training is required for those who specify glove types within a permit to work system.

M. General Considerations

- All contact with chemicals should be minimized, and if significant contamination occurs, the gloves should be washed as soon as possible. Grossly contaminated gloves should not be reused and if a user is in doubt regarding the state of a pair of gloves then they should obtain new ones.
- Factors such as grip and physical strength of the glove must be considered in the selection process.
- For dry powders any "chemical resistant" glove can be used.
- With mixtures and formulated products, unless specific tests data are available a glove should be selected on the basis of the component with the shortest breakthrough time as it is possible for solvents to carry active ingredients through polymeric materials.
- Individuals should not be allowed to eat, drink, or smoke while wearing gloves.

- As long as the performance characteristics are acceptable, in certain circumstances it may be more cost effective to regularly change cheaper gloves than to reuse more expensive types.
- Sleeves should be worn over the cuffs of gloves. There should be no gap between the sleeve and the glove.
- Care should be taken to ensure that only clean hands are put into gloves. Chemical contamination of the internal surfaces of gloves means that the substances will be held in contact with the skin. This increases the potential for absorption through the skin.
- Care should be taken not to spread contamination when wearing gloves, especially onto items that people not wearing gloves may touch. In addition people should be careful about touching unprotected skin with contaminated gloves.
- Certain individuals can suffer a skin reaction to a constituent of the material of construction of a particular type of glove.

VII. CONCLUSIONS

Gloves have an important role to play in providing adequate protection in the workplace. However, the benefit will not be fully realized unless there are good procedures for selection, issue, use, cleaning, storage, and training, and management is committed to ensuring the gloves are used at all times they are required.

The above scheme for the selection and use of gloves, on the basis of standardized test data, is intended to provide guidance rather than to be adhered to strictly. When used in conjunction with local knowledge and experience of the work undertaken and the chemicals used, and consideration of other factors, such as the standards of training and supervision, it should assist in the development of rational rules and procedures for the selection and use of gloves.

Although not all of the information may be available to enable an individual to be completely certain that the choice of glove and the associated procedures in a given situation will provide the protection desired, if the guidelines outlined in this chapter are followed the care taken in the process will provide the users with more protection than is often afforded.

REFERENCES

1. HMSO, Management of Health and Safety at Work Regulations, 1992. Approved Code of Practice, Her Majesty's Stationery Office, London, 1992.
2. HMSO, Personal Protective Equipment at Work Regulations, 1992. Approved Code of Practice, Her Majesty's Stationery Office, London, 1992.
3. Leinster, P., Bonsall, J.L., Evans, M.J., and Lewis, S.J., The application of test data in the selection and use of gloves against chemicals, *Ann. Occup. Hyg.*, 34, 85, 1990.

4. HMSO, The Classification, Packaging and Labelling of Dangerous Substances Regulations 1984, plus amendments, Her Majesty's Stationery Office, London.
5. British Occupational Hygiene Society Technology Committee Working Party on Protective Clothing, The development of a standard test method for determining permeation of liquid chemicals through protective clothing materials, *Ann. Occup. Hyg.*, 30, 381, 1986.
6. ASTM, *Standard Test Method for Resistance of Protective Clothing Materials to Permeation by Hazardous Liquid Chemicals, F739-81*, American Society for Testing and Materials, Philadelphia, USA.
7. BS1561, Specification for Industrial Gloves, British Standards Institution, London, 1986.

20

The Selection and Use of Gloves By Health Care Professionals

Lars G. Burman and Birgitta Fryklund

TABLE OF CONTENTS

    I. Introduction 283
       A. Historical Aspects 283
       B. Quality vs. Cost 284
   II. The Use of Gloves in Health Care 285
       A. General Aspects 285
          1. Rationale for Gloving of HCWs 285
          2. Basic Glove Usage Routines 285
       B. Categories of Gloves 286
       C. Selection of Gloves and Importance of Guidelines 286
       D. Compliance With Glove Usage Guidelines 286
       E. Storage of Gloves 288
  III. Effectiveness of Gloves for Prevention of Transmission of Infection 288
  IV. Limitations of Gloves as a Barrier to Microorganisms 289
       A. Imperfect Manufacturing 289
       B. Glove Perforation 289
   V. Improving the Effectiveness of Gloves 289
       A. Double-Gloving 289
       B. Extra Protective Gloves 290
References 290

I. INTRODUCTION

A. Historical Aspects

The use of gloves for protection of staff and patients has a long tradition in medicine. The surgical glove was introduced in 1889 by the neurosurgeon William Stuart Halsted (1852–1922) at Johns Hopkins University School of

Medicine in Baltimore, U.S. His favorite scrub nurse, Caroline Hampton, had developed an allergy to mercury chloride, the chemical disinfectant used during the surgical procedures. The Goodyear Company was therefore given the assignment of manufacturing a reusable latex glove for protection of the nurse's hands during operations. This deed by the Baltimore surgeon may have contributed to his successful proposal to later marry Miss Hampton.

The invention of the surgical glove marked the transition between Lister's antisepsis and the asepsis era in surgery. Around the turn of the century the use of sterile latex gloves during surgery had become accepted as a complement to the preoperative hand disinfection, and sterile sutures, towels, dressings, and gowns as part of the efforts to prevent surgical infections.[1] Since then, the use of gloves during different diagnostic and patient care procedures in health care has gradually been widened. The evolution of manufacturing processes, materials, and designs of gloves and, in particular, glove usage, has nearly exploded in recent years fuelled by the risk of occupational blood-borne human immunodeficiency virus (HIV) infection. Thus, the original reusable latex glove has been replaced by a large number of types of single-use gloves adapted for various tasks, from protection of patients and staff during minute-long minor procedures to heavy surgical procedures with a duration of many hours.

B. Quality vs. Cost

The main purpose of using gloves in the health care setting is thus no longer to protect the staff from allergy, but to reduce transmission of viruses, bacteria, and fungi among patients and between patients and health care workers in order to minimize the risk of nosocomial infection. Although at least 95% of these infections are caused by bacteria, the HIV pandemic and the incurable nature of the disease has markedly increased the interest in the quality of glove materials, standards, and testing of quality.

The so-called universal precautions policy recommended by the U.S. Centers for Disease Control (CDC) calls for wearing gloves whenever a health care worker (HCW) expects to get in contact with mucosal surfaces, blood, or certain other body secretions from a patient.[2] The increasing use of gloves in health care, including dentistry, is not only due to the rising numbers of HIV-infected patients, but also to protect other patients with compromised immune defense due to extreme age, various invasive procedures, and therapies involving immunosuppression. In these cases, the gloves worn by HCWs may reduce the risk of colonization and infection by exogenous microorganisms. To illustrate the sudden increase in the cost of gloves, a recent U.S. study showed that a 900-bed hospital spent $180,000 per annum on gloves excluding those used in operating theaters.[3]

Because of the rapidly increasing usage of gloves by HCWs and other professionals (ambulance staff, police, etc.) questions concerning appropriate guidelines (indications for using gloves, the proper selection of gloves of different quality for different tasks, compliance to policies among HCWs), the price vs. quality of glove materials and their propensity to cause allergy have become increasingly important.

II. THE USE OF GLOVES IN HEALTH CARE

A. General Aspects

1. Rationale for Gloving of HCWs

There are three distinct theoretical reasons for HCWs to wear gloves. First, gloves reduce the risk of personnel infection due to microorganisms from infected patients. For example, gloves should prevent the nursing staff from developing herpetic whitlow after giving oral care or suctioning a patient with oral *Herpes simplex* infection. In particular, one reported case of acquisition of HIV infection by a HCW through soiling of chapped hand skin with patient blood has caught much attention.[4]

Second, gloves reduce the likelihood of personnel transmitting their own endogenous microbial flora to patients. For example, sterile gloves are worn by the personnel for this reason when performing operations as well as when they need to touch large open wounds, e.g., during dressing of burn wounds.

Third, gloves reduce the risk that personnel will become transient carriers of microorganisms from one patient and later transmit these to other patients. Under most conditions, such transient skin colonization of the hands can be eliminated by hand washing, but this policy may be insufficient with highly contagious infectious agents such as respiratory syncytial virus (RSV) infection, or rotavirus enteritis in the neonate, or antibiotic associated *Clostridium difficile* diarrhea in elderly hospitalized patients.

Thus, in hospitals where hand washing is performed carefully and appropriately by all staff members, gloves are theoretically seldom necessary to prevent transient colonization of staff members and subsequent transmission by them to others. However, since numerous studies show that hand washing practices are inadequate in most hospitals, even when the staff is advised and surveilled for study purposes,[5-7] gloves represent the most efficient way of preventing transient hand colonization and transmission of microorganisms in order to reduce the risk of nosocomial infections. Wearing gloves is always indicated for touching the patients' excretions, secretions, blood, or other body fluids, because these are generally regarded as potentially infectious material. Gloves may not be needed if "no touch" technique (not touching infectious material with hands) can be used, e.g., during dressing of minor wounds.

2. Basic Glove Usage Routines

When gloves are indicated or recommended, disposable single-use gloves (sterile or nonsterile and of varying quality, depending on the purpose for use, see below) should be worn. Used gloves should be discarded into an appropriate receptacle.

Gloves must of course be discarded and the hands washed and/or disinfected between patients. Hand washing is important because no glove is 100% impermeable to microorganisms (as described in Chapter 9). Washing of gloves between patients is not recommended because compliance will probably be even lower than for hand washing. Also, disinfection of gloves is not recommended because of the risk of destroying the glove's protective capacity. After direct contact with a patient's excretions or secretions when taking care of that patient, gloves should be changed even if care of that patient has not been completed.

B. Categories of Gloves

From a practical point of view, gloves used by HCWs, ambulance staff, police, etc., can be divided into three major categories.

Surgical glove — Sterile, powdered by various methods and made of latex for optimal compromise between elasticity, tensile strength, fit, and tactility. Surgical gloves made of expensive special synthetic materials are available for HCWs with latex allergy.

Examination glove — Usually nonsterile, made of latex or polyvinylchloride (PVC), and with moderate fit (particularly true for vinyl gloves). For special diagnostic, therapeutic, or other patient care procedures and dentistry.

Protective glove — Nonsterile, inexpensive, made of, e.g., polyethylene. Used for short-term, "dirty" patient care procedures, such as handling of urine and feces.

C. Selection of Gloves and Importance of Guidelines

A rational use of gloves in a hospital requires a carefully prepared policy in writing. It should be laid down by, e.g., representatives of the nursing staff, the infection control unit, and the purchasing department. It is also important to keep all concerned personnel well informed about the guidelines and to receive feedback from the staff so that necessary adjustments of the guidelines can be made. Another important aspect of a glove policy is to have the appropriate types and sizes of gloves available where and when they are needed. Lacking glove usage guidelines implies microbiological hazards to patients and staff and may lead to considerable extra costs because of overusage of gloves or selection of gloves of unnecessarily high quality for the task.

A simplified example of glove usage guidelines for medical and dental care is shown in Table 1.

D. Compliance With Glove Usage Guidelines

The importance of proper glove usage by hospital staff was poignantly underscored by the report on a case of non-needle stick occupational

TABLE 1 Selection of Glove in Different Situations

Purpose	Procedure	Type of glove	Infection or microorganism
Protection of personnel	Dentistry	Examination, non sterile	Blood-borne: hepatitis (A), B, C, HIV, HTLV[c]
	Surgery	Surgical	(Hemorrhagic fevers)
	Risk of contact with blood	Protective[a] Examination, nonsterile[b]	
Protection of personnel and patients	Handling of feces, urine, vomit, etc.	Protective	Various viruses and bacteria[d]
Protection of patients	Dentistry	Examination, nonsterile	Hepatitis HIV
	Surgery	Surgical	Staphylococci, hepatitis, HIV
	Other invasive procedures[e]	Examination, sterile	Staphylococci
	Isolation Barrier nursing	Examination or protective	Various viruses and bacteria[d]
	Handling of feces, urine, etc.	Protective	Special viruses and bacteria[f]

[a] e.g., Wiping of blood spillage.
[b] Phlebotomy.
[c] Human lymphotropic virus I and II.
[d] e.g., Rotavirus, *Salmonella* spp, other Gram negative bacteria, *Clostridium difficile*, *Staphylococcus aureus*, yeasts.
[e] e.g., Initiation and maintenance of central venous catheter.
[f] e.g., Rotavirus, respiratory syncytial virus, staphylococci, *C. difficile*.

transmission of HIV.[4] In spite of this, highly variable compliance with glove usage guidelines has been reported. In one recent study, compliance with CDC guidelines for performance of procedures that may involve contact with body fluids, particularly blood, was 92% for arterial blood gas procedures, 78% for i.v. line initiation and maintenance, and 71% for phlebotomy.[3]

In another recent study, the investigators judged the rate of appropriate glove usage to be 90% in the laboratory, but only 59% on the wards.[8] Approximately half of the gloves used on the hospital wards in this study were being wasted because they were not considered necessary according to the hospital's policy. On removing gloves, only 52% of the nurses washed their hands.

Other studies have also found a rather low degree of compliance with different aspects of universal precaution policies such as glove usage.[9] Thus, there seems to be much room for improved compliance with the glove usage guidelines in most hospitals.

E. Storage of Gloves

Because the quality of the gloves, and the ward and product packages which keep the gloves sterile until use may deteriorate due to a variety of physical agents, the storage of gloves in hospitals is important. The following guidelines represent major points excerpted from British Standard.[10]

1. The gloves should be kept in their transportation or ward package until used.
2. The temperature should be below 25°C and the gloves should not be stored near any source of heat nor in direct sunlight.
3. The relative humidity of the air may not be so high that there is condensation.
4. Sources of light: the gloves should be stored in darkness, protected from the sun and the light from fluorescent tubes.
5. Radiation: the gloves should not be stored near any source yielding ionized radiation, e.g., X-ray apparatus.

III. EFFECTIVENESS OF GLOVES FOR PREVENTION OF TRANSMISSION OF INFECTION

The role of gloves in prevention of surgical infections, regrettably, escapes scientific evaluation. When gloves were introduced into surgical practice at the turn of the century, our current standard method for evaluation of therapies — the randomized clinical trial — had not yet been developed. Thus, the lowered postoperative infection rates noted were related to historical controls. In addition, gloves were introduced together with other new protective measures (sterile sutures, towels, dressings, operating gowns, etc.) which prevented an evaluation of the contribution by each component to the markedly reduced incidence of infection.[1] Today, a randomized study of sterile gloves vs. bare hands in surgery would be impossible for a number of reasons (unethical both to patients and operating staff).

A number of retrospective, and thus inconclusive, studies attempting to establish an impact of gloves on hepatitis B virus (HBV) transmission from oral surgeons or dentists to their patients or vice versa have been negative.[11-14] On the other hand, studies in renal dialysis units have documented that the routine use of gloves in conjunction with other infection-control procedures can substantially reduce the excess risk of HBV infection among HCWs.[15] Also, several prospective, competently designed, well-executed studies have shown a superiority of gloves to regular hand washing as part of barrier nursing in special situations.[16,17] Thus, at least for agents with high transmissibility in the hospital setting, such as RSV and various bacteria in neonatal intensive care units and *C. difficile,* gloves seem to be essential for preventing transmission of infection.

IV. LIMITATIONS OF GLOVES AS A BARRIER TO MICROORGANISMS

A. Imperfect Manufacturing

Neither vinyl nor latex gloves may be regarded as completely impermeable to microorganisms, both because of the porosity of the material and the varying proportion of gloves with pinholes as a result of imperfect fabrication processes (see Chapters 3 and 9). Although one study found one type of latex glove impermeable to HIV,[18] several other studies have shown that synthetic latex, Neoprene®, vinyl, etc., allow penetration of various viruses.[19-22] Therefore, disposable gloves shall be replaced as soon as practical when contaminated, but obviously some critical procedures (e.g., surgery) cannot always be interrupted in order to change the gloves of the operating team. This is, however, often routinely done, e.g., after skin incision in hip arthroplasty or after removing the excised part of the bowel in colonic surgery. As mentioned, disposable single-use gloves must not be washed or disinfected and reused.

The problem of pinholes due to imperfect manufacturing is to certain degrees accepted by current quality standards and has been documented by many reports. In one study, 0 to 85% of disposable examination gloves from 17 different manufacturers were defective due to holes, and 0 to 18% of sterile surgical gloves from one manufacturer, depending on the batch.[23]

B. Glove Perforation

Numerous studies have shown a substantial rate of glove perforations during surgical procedures. In one recent report from a neurosurgical unit 43% of 550 gloves collected from 100 consecutive operations had been damaged.[24] In another recent study of 2292 general surgical procedures, major glove tears occurred in 11%. In two thirds of these there was visible skin contact with the patient's blood, and in 3% there occurred sharp injuries of operating team members.[25] However, in a hospital glove puncture, accidents due to cuts, and above all, needle sticks, are much more common outside operating theaters than various sharp injuries occurring during surgery.[26]

V. IMPROVING THE EFFECTIVENESS OF GLOVES

A. Double Gloving

Double-gloving has been shown to reduce the glove perforation rate in surgery (e.g., Matta et al.[27]). In the latter study the rate of puncture of the outer glove was 11% and of the inner glove 2%, suggesting protection in four out of five cases in which the outer glove had been breached. Indeed, double-gloving apparently reduces the risk of HBV transmission from surgeon to patient.[28]

Thus, double-gloving is advisable, e.g., when performing surgical procedures on known HIV or HBV positive patients or when the infection risk is very high such as in implant surgery.

B. Extra Protective Gloves

Gloves that are more resistant to scalpel laceration (cut resistant) than latex gloves are now becoming available to surgical personnel. However, needle sticks were the most common injuries incurred by surgeons in a recent study of 2106 operations.[29] In this study, 107 of 112 accidental injuries reported were due to needle sticks, four were ascribed to scalpel cuts, and one was the result of diathermy burn. No commercially available needle stick-proof gloves have yet been developed because of difficulties in finding materials with enough puncture resistance.

One alternative to reduce the risk of accidental needle stick infection by viruses such as HIV may be a multilayered "hybrid" glove that can resist needle-sticks in selected areas of increased armor and reduce or prevent infection by chemically inactivating viruses when needles penetrate the glove. In a newly published study, glove materials containing chlorhexidine gluconate in an instant-release matrix on their inner surfaces were able to rapidly inactivate various bacteria and viruses, including a model retrovirus (HIV-like) and perhaps HBV in a bench test.[30] In another recent laboratory study, combinations of glove materials were tested for reduction of transfer of HIV-1 to cell cultures by needle puncture.[31] The rate of HIV-1 infection of cell cultures after glove puncture was greater than 90% with a single latex surgical glove barrier, 60 and 23% with double or triple layers of latex, respectively, less than 8% with an intermediate cotton layer impregnated with 4% of the antimicrobial and spermicidal compound nonoxynol-9, 6% with an intermediate layer made of Kevlar® (a fabric of olefin polyester fibers of extreme physical strength), and 0% with an intermediate Kevlar® layer impregnated with nonoxynol-9.

Gloves made of these new materials with improved resistance to needle sticks and scalpel cuts and containing antimicrobial substances may prove to further reduce the risk of transmission of blood-borne pathogens between patients and hospital personnel in the future.

REFERENCES

1. Brewer, G. E., Studies in aseptic technique, *J. Am. Med. Assoc.*, 64, 1369, 1915.
2. Centers for Disease Control, Update: universal precautions for prevention of transmission of human immunodeficiency virus, hepatitis B virus, and other bloodborne pathogens in health care settings, *MMWR*, 37, 377, 1988.
3. Stringer, B., Smith, J. A., Scharf, S., Valentine, A., and Walker, M. M., A study of the use of gloves in a large teaching hospital, *Am. J. Infect. Control*, 19, 233, 1991.

4. Centers for Disease Control, Update: human immunodeficiency virus infections in health-care workers exposed to blood of infected patients, *MMWR*, 36, 285, 1987.
5. Albert, R. K. and Condie, F., Handwashing patterns in medical intensive-care units, *N. Engl. J. Med.*, 304, 1465, 1981.
6. Doebbeling, B. N., Stanley, G. L., Sheets, C. T., Pfaller, M. A., Houston, A. K., Annis, L., Li, N., and Wenzel, R. P., Comparative efficacy of alternative handwashing agents in reducing nosocomial infections in intensive care units, *N. Engl. J. Med.*, 327, 88, 1992.
7. Graham, M., Frequency and duration of handwashing in an intensive care unit, *Am. J. Infect. Control*, 18, 77, 1990.
8. Kaczmarek, R. G., Moore, R. M., McCrohan, J., Arrowsmith-Lowe, J. T., Caquelin, C., Reynolds, C., and Israel, E., Glove use by health care workers. Results of a tristate investigation, *Am. J. Infect. Control*, 19, 228, 1991.
9. Kelen, G. D., Fritz, S., Qaquish, B., Brookmeyer, R., Baker, J. L., Kline, R. L., Cuddy, R. M., Goessel, T. K., Floccare, D., Williams, K. A., Siverston, K. T., Altman, S., and Quinn, T. L., Unrecognized human immunodeficiency virus infection in emergency department patients, *N. Engl. J. Med.*, 318, 1645, 1988.
10. British Standard (BS 3574:1989), Her Majesty's Stationery Office, London.
11. Mosley, J. W., Edwards, V. M., Casey, G., Redeker, A. G., and White, E., Hepatitis B virus infection in dentists, *N. Engl. J. Med.*, 293, 729, 1975.
12. Weil, R. B., Lyman, D. O., Jackson, R. J., and Bernstein, B., A hepatitis sero-survey of New York dentists, *N.Y. State Dent. J.*, 43, 587, 1977.
13. Reingold, A. L., Kane, M. A., and Hightower, A. W., Failure of gloves and other protective devises to prevent transmission of hepatitis B virus to oral surgeons, *J. Am. Med. Assoc.*, 259, 2558, 1988.
14. Noble, M. A., Mathias, R. G., Gibson, G. B., and Epstein, J. B., Hepatitis B and HIV infections in dental professionals: effectiveness of infection control procedures, *J. Can. Dent. Assoc.*, 57, 55, 1991.
15. Snydman, D. R., Bryan, J. A., Macon, E. J., and Gregg, M. B., Hemodialysis-associated hepatitis. Report of an epidemic with further evidence on mechanisms of transmission, *Am. J. Epidemiol.*, 104, 563, 1976.
16. Leclair, J. M., Freeman, J., Sullivan, B. F., Crowley, C. M., and Goldmann, D. A., Prevention of nosocomial respiratory syncytical virus infections through compliance with glove and gown isolation precautions, *N. Engl. J. Med.*, 317, 329, 1987.
17. Klein, B. S., Perloff, W. H., and Maki, D. G., Reduction of nosocomial infection during pediatric intensive care by protective isolation, *N. Engl. J. Med.*, 320, 1714, 1989.
18. Dalgleish, A. G. and Malkowsky, M., Surgical gloves as a mechanical barrier against human immunodeficiency viruses, *Br. J. Surg.*, 75, 171, 1988.
19. Gerhardt, G. G., Results of microbiological investigations on the permeability of procedure and surgical gloves, *Zbl. Hyg.*, 188, 336, 1989.
20. Korniewiez, D. M., Langhon, B. E., Cyr, W. H., Lytle, C. D., and Larson, E., Leakage of virus through used vinyl and latex examination gloves, *J. Clin. Microbiol.*, 28, 787, 1990.

21. Marin, J., Dragas, A. Z., and Mavsar, B., Virus permeability of protective gloves used in medical practice, *Zbl. Hyg.,* 191, 516, 1991.
22. Kotilanen, H. R., Avato, J. L., and Gantz, N. M., Latex and vinyl nonsterile examination gloves, status report on laboratory evaluation of defects by physical and biological methods, *Appl. Environ. Microbiol.,* 56, 1627, 1990.
23. Daschner, F. D. and Habel, H., HIV prophylaxis with punctured gloves (letter), *Infect. Control Hosp. Epidemiol.,* 9, 184, 1988.
24. Palmer, J. D. and Rickett, J. W. S., The mechanisms and risks of surgical glove perforation, *J. Hosp. Infect.,* 22, 279, 1992.
25. Wright, J. G., McGeer, A. J., Chyatte, D., and Ransohoff, D. F., Mechanisms of glove tears and sharp injuries among surgical personnel, *J. Am. Med. Assoc.,* 266, 1668, 1991.
26. Shanson, D., The risks of transmission of the HTLV-III and hepatitis B virus in the hospital, *Infect. Control,* 7, 128, 1986.
27. Matta, H., Thompson, A. M., and Rainey, J. B., Does wearing two pairs of gloves protect operating theatre staff from skin contamination?, *Br. Med. J.,* 297, 597, 1988.
28. Carl, M., Francis, D. P., Blakey, D. L., and Maynard, J. E., Interruption of hepatitis B transmission by modification of a gynaecologist's surgical technique, *Lancet,* 1, 731, 1982.
29. Hussain, S. A., Latif, A. B. A., and Choudhary, A., Risk to surgeons. A survey of accidental injuries during operations, *Br. J. Surg.,* 75, 314, 1988.
30. Modak, S., Sampath, L., Miller, H. S. S., and Millman, J., Rapid inactivation of infectious pathogens by chlorhexidine-coated gloves, *Infect. Control Hosp. Epidemiol.,* 13, 463, 1992.
31. Johnson, G. K., Nolan, T., Wuh, H. C., and Robinson, W. S., Efficacy of glove combinations in reducing cell culture infection after glove puncture with needles contaminated with Human Immunodeficiency Virus type 1, *Infect. Control Hosp. Epidemiol.,* 12, 435, 1991.

21

Information Sources for Glove Test Data

Gunh A. Mellström

TABLE OF CONTENTS

I. Degradation and Permeation Test Data 293
    A. Printed Reports 294
        1. Journal Articles 294
        2. Test Reports 294
        3. Symposium and Congress Publications 294
    B. Chemical Resistance Guiding Lists 295
    C. Databases 296
    D. Guidelines for the Selection of Gloves 298
II. Penetration Test Data 298
    A. Printed Reports 298
    B. Quality Control Test Reports 299
III. Dermatological Side Effect Reports on Gloves 299
    A. Printed Reports 299
    B. Databases 300
References 300

I. DEGRADATION AND PERMEATION TEST DATA

When chemical resistance for protective gloves are presented as degradation test data it can be reported as mean weight or mean thickness changes, but also by rating the visual changes using a given scale.[1]

Rating of the material	Visual changes after
1	5 min
2	30 min
3	1 h
4	4 h

Chemical protective materials (gloves) that receive a 1 and 2 rating or more than 20% weight or thickness change are considered nonacceptable. Chemical resistance has also been rated by other scales after visual inspection for example, excellent, good, fair, or not recommended. This kind of resistance information is still occurring in certain manufacturers' chemical resistance lists. Today most test data are from permeation testing and presented as breakthrough time (min) and permeation rates ($\mu g/m^2$ min) and can be found in different kind of sources.

A. Printed Reports

1. Journal Articles

Primary test data from permeation and degradation testing are often presented in scientific and/or technical journals with stress on industrial hygiene or environmental research such as:

> American Industrial Hygiene Association Journal
> Applied Industrial Hygiene
> Annals of Occupation Hygiene
> Environmental Research
> International Archives of Occupational Environmental Health
> Occupational Health and Safety
> Travail et Sécurité

2. Test Reports

This kind of test data can also be published as reports from research and test laboratories or institutes, for example:[2-7]

> Arbetsmiljöfondet, Denmark
> Lawrence Livermore Laboratory
> Radian Corporation
> National Institute of Occupational Health, Sweden
> Texas Research Institute
> U.S. Coast Guard
> National Institute of Occupational Safety and Health, U.S.

3. Symposium and Congress Publications

Papers presented at international congresses and symposia are often published as proceedings. The ASTM Committee F-23 on Protective Clothing has sponsored four international symposia on the performance of protective clothing. One of general subjects of these symposia has been chemical protection and includes test methodology for evaluating the permeation resistance of protective clothing materials, field evaluation methods for end-use items of protective clothing, decontamination techniques, and risk assessment in the selection and use of chemical protective clothing. The first international symposium was held in Raleigh, North Carolina, in 1984 and the papers presented at this symposium were published in the ASTM, STP 900 (Special Technical Publication) and contains 48 papers.[8] The second international symposium was held in Tampa, Florida, in 1987 and the papers presented at this symposium were published in the ASTM, STP 989, which contains 84 papers selected from the original presentations.[9] The third symposium was held in San Diego, California in 1989, and provided a review of the problems, new technologies, and uses of protective clothing related specifically to chemical emergency response. The papers presented at this symposium were published in the ASTM, STP 1037 and contains 21 papers.[10] The fourth international symposium was held in Montreal, Canada in 1991 and 80 papers from this symposium are published in ASTM, STP 1133.[11]

A permanent Scandinavian association was established in 1984 at a symposium in Copenhagen with the name NOKOBETEF, NOrdisk KOordinationsgruppe om BEskyttelseklaeder som TEknisk Forebyggelsemiddel, (The Nordic Coordination Group on Protective Clothing as a Technical Prophylactic Measure). A second meeting was held in Stockholm in 1986, this time also with participants from countries outside Scandinavia. The third international meeting was held in Gausdal in Norway and the fourth international meeting was held in Kittilä, Finland in 1992. Proceedings from the NOKOBETEF meetings have been published and can be ordered from the NOKOBETEF c/o Hazprevent, Dronningensgade 25, DK-1420 Copenhagen, Denmark. The second and third symposia had as the general subject protective clothing against chemicals and other health risks and contains 21 abstracts plus 8 posters, and 36 abstracts plus 4 poster presentations, respectively.[12-14] The fourth meeting addressed the quality and usage of protective clothing and 43 papers plus 8 posters were presented.[15]

Papers with test data such as breakthrough time and permeation rate from permeation testing of protective gloves, published in the ASTM STP publications and in proceedings from NOKOBETEF meetings have been put into the database DAISY, described below.

B. Chemical Resistance Guiding Lists

The largest manufacturers of protective gloves list their permeation test results for their protective products.[16-22] The test results are usually from standard permeation testing performed at their own test control laboratories or at well recognized test laboratories, giving breakthrough time values and sometimes also the permeation rate for the chemicals tested through the glove materials. These kinds of test results are, as a rule, included in the databases described below.

C. Databases

Name:	CPCbase[23]
Producer:	A. D. Little, Inc., Cambridge MA, U.S.
Hardware:	IBM or IBM compatible personal computer (PC)
Software:	Data flex
Commercial facilities:	Diskettes, commercially available
Data stored:	Results from permeation and immersion testing as breakthrough time, permeation rate, weigh-changes, description of the clothing materials tested (gloves included), their sources and literature references.

Name:	DAISY, a reference database on protective gloves for occupational use[24]
Producer:	National Institute of Occupational Health, Department of Occupational Dermatology, Solna, Sweden
Hardware:	Macintosh
Software:	4th Dimension
Commercial facilities:	Available by a questionnaire service at the National Institute of Occupational Health, Department of Occupational Dermatology, Phone: 46 8 730 93 25, Fax: 46 8 730 98 92. The information is in English.
Data stored:	Information on protective effects of gloves made of rubber and plastic against chemicals. The database contains information on primary data from material permeation testing such as chemical name and CAS nr, breakthrough time, permeation rate, material thickness, product name, and description. It covers more than 500 chemicals or aqueous solutions. This database also contains test data from experimental permeation testing of glove materials intended to reduce the percutaneous absorption and from clinical tests on patients, protective effects as well as side effects. Bibliographic data of the test reports are given. Continuously updated.

Name:	Guidelines for the Selection of Chemical Protective Clothing[1]
Location:	EPA, Washington D.C., U.S.
Producer:	A. D. Little, Inc., Cambridge, MA, U.S.
Hardware:	Mainframe computer at EPA
Software:	Own
Commercial facilities:	Accessible through an on-line system
Data stored:	Permeation and degradation primary data on protective clothing materials, gloves included. It contains permeation data for approximately 500 chemicals or aqueous solutions, 25 multicomponents organic solutions and 12 major clothing materials, gloves included.

Information Sources for Glove Test Data 297

Name:	Chemical Protective Clothing Permeation/Degradation Database — IBM version
Producer:	Forsberg, K., A.G.A.-AB, Lidingö, Sweden, Keith, L. Radian Corporation, Austin, TX, U.S.
Hardware:	IBM or IBM compatible PC, a hard drive with 1 MB available
Software:	DOS Version 2.0 or higher
Commercial facilities:	Diskettes, commercially available from Lewis Publishers, Boca Raton, FL, U.S.
Data stored:	Name of chemical compound or mixtures tested, material tested (gloves included), manufacturer, product name, thickness, breakthrough times, permeation rate and reference source. 12,000 test results of more than 660 chemicals against 250 Chemical Protective Clothing (CPC), gloves included. Periodically updated. Equal to the Macintosh version.

Name:	Chemical Protective Clothing Permeation/Degradation database — Macintosh version
Producer:	Blotzer, M., Forsberg, K., A.G.A.-AB, Lidingö, Sweden, Keith, L. Radian Corporation, Austin, TX, U.S.
Hardware:	Macintosh Plus, SE, Portable, or any Macintosh II computer with 2MB or more RAM and a hard drive.
Software:	Hyper Card Version 2.0 or higher
Commercial facilities:	Diskettes, commercially available from Lewis Publishers, Boca Raton, FL, U.S.
Data stored:	Name of chemical compound or mixtures tested, material tested (gloves included), manufacturer, product name, thickness, breakthrough times, permeation rate and reference source. 12,000 test results of more than 660 chemicals against 250 Chemical Protective Clothing (CPC), gloves included. Periodically updated. Equal to the IBM version.

Name:	GlovES+, Expert System[25]
Location:	Keith L. Radian Corporation, Austin, TX, U.S.
Producer:	Keith L. Radian Corporation, Austin, TX, U.S.
Hardware:	IBM or IBM compatible PC
Software:	Ruelemaster
Commercial facilities:	Diskettes, commercially available
Data stored:	The permeation test data is from CPC Performance Index, but defined rules are applied to provide data applicable to specific situations using a menu system. Data on products that would not meet the specific requirements is not presented on the screen. No bibliographic data are given.

D. Guidelines for the Selection of Gloves

The guidelines presented below can be used for the selection of protective gloves even if some of the protective materials or some qualities of the materials are not available as gloves. Usually the materials are listed and those used for gloves are indicated.

Guidelines for the Selection of Chemical Protective Clothing, 3rd ed., Vol. I and II is prepared by A. D. Little, Inc., Cambridge, MA and published by ACGIH, Inc., Cincinnati, OH.[26] Volume I contains discussions of the basic concept of permeation and chemical resistance and test methods for CPC. It also contains two matrices for the selection of protective clothing which presents clothing recommendations for 12 major clothing materials (gloves included) and about 500 chemicals or aqueous solutions. Volume II contains a detailed discussion of permeation theory and testing methods and could be considered a supporting document to Volume I.

Chemical Protective Clothing Performance Index Book[25] contains a presentation of the performance index numbers, the GlovES+ Expert system, chemical class index, and risk codes for chemicals. It also contains reported permeation test data for about 650 chemical compounds and mixtures, and references from which the data was obtained up to 1988. An updated edition is in preparation.

The intention of the *Quick Selection Guide to Chemical Protective Clothing*[27] is to assist workers, supervisors, safety and health personnel, to select the best protective clothing materials for the job. The guide is only a summary of data and should not be the single basis for the final selection of protective clothing or protective gloves. Color-coded tables summarize the chemical breakthrough performance of 11 generic materials against approximately 450 chemicals. It is periodically updated.

Comprehensive information on chemical protective clothing, gloves included, can also be found in: *Chemical Protective Clothing,* Vol. I and II, by J. S. Johnson and K. J. Anderson,[28] and *The National Toxicology Program's Chemicals Data Compendium,* Volume VI, Personal Protective Equipment, by L. C. Keith and D. B. Walters.[29]

II. PENETRATION TEST DATA

A. Printed Reports

The penetration standard test methods are used to investigate the penetration of chemicals and/or microorganisms through imperfections in the glove material. Test results from this kind of investigation are usually presented in articles in appropriate scientific journals and in congress/symposium books. As the penetration of both chemicals and microorganisms are concerned, the results are presented in journals covering different fields, for example:[30-45]

Applied and Environmental Microbiology
Archives of International Medicine
ASTM STP 1133
Biotechniques
British Journal of Surgery
British Medical Journal
HYGIENE + Medizin, Aus der Praxis
Infection Control Hospital Epidemiology
Journal of Acquired Immune Deficiency Syndrome
Journal of American Dental Association (JADA)
Journal of American Medical Association (JAMA)
Journal of Clinical Engineering
Journal of Clinical Microbiology
Journal of Hospital Infection
Nursing Research

B. Quality Control Test Reports

Standard penetration test methods are also used as quality control testing for freedom from holes in gloves. This kind of test data can be collected and presented in reports for internal and external use. For example: *Glove Update*, (1989) is a FDA report summarizing test results from a 1000 ml leak test. An edited copy of this report can be requested from Fred Sadler, The Freedom of Information Office HFZ-82, 5600 Fisher's Lane, Rockville, MD 20857;[46] *SPRIMA-Test Examination Gloves*, SEMKO AB, Medical devices, Stockholm, Sweden (in Swedish);[47] and *SPRIMA-Test Surgical Gloves*, SEMKO AB, Medical devices, Stockholm, Sweden (in Swedish).[48]

To our knowledge there are no computerized databases commercially available today containing these kinds of test data.

III. DERMATOLOGICAL SIDE EFFECT REPORTS ON GLOVES

A. Printed Reports

Most side effect reactions when using rubber and plastic gloves are reported in dermatological/medical journals as case reports or from investigations of patients with a certain occupation where gloves are commonly used, such as hospital personnel and among house cleaning employees. Examples of such journals are

Allergy
British Journal of Dermatology
British Medical Journal
Contact Dermatitis
Dermatosen
Hautartz

Journal of American Dental Association (JADA)
The New England Journal of Medicine

Contact dermatitis from rubber gloves caused by the rubber chemicals have been well known for quite some time and are well documented. In most dermatological textbooks published in recent years there are chapters dealing with side effects caused by rubber gloves and other rubber products. Some of the most well known are *Contact Dermatitis,* by E. Cronin;[49] *Occupational Skin Diseases,* by R. M. Adams;[50] *Occupational and Industrial Dermatology,* by H. I. Maibach, Ed.;[51] *Textbook in Contact Dermatitis,* by R. J. G. Rycroft, T. Menné, P. J. Frosch, and C. Benezra, Eds.;[52] and *Toxic Hazards of Rubber Chemicals,* by A. R. Nutt.[53]

In recent years there have also been some research reports published as thesis in this matter such as *Latex Glove Contact Urticaria,* by K. Turjanmaa,[54] and *Occupational Skin Diseases in Finland,* by T. Estlander.[55]

The number of adverse reactions to latex medical devices, among these, latex gloves, has grown the recent years and resulted in the first international conference on latex reactions: Sensitivity to Latex in Medical Devices, November 5 through 7, 1992, sponsored by the Food and Drug Administration, Centers for Disease Control and the National Institute of Allergy and Infectious Diseases. More than 40 papers were presented in sessions for basic studies, clinical studies, practical and manufacturing approaches and another 15 presentations in a poster session.[56]

C. Databases

Up until now, DAISY, discussed previously, is the only database that contains specific test data on side effect reactions caused by gloves for occupational use. The side effect reaction data has been collected from the journal articles, congress and symposia papers previously mentioned. The data stored are in addition to chemical permeation data, case reports on contact dermatitis and contact urticaria occurring when working with gloves, and also patch and prick test reactions from gloves tested on patients.

On the other hand, there are several databases in the dermatological field containing test reactions from different points of view. Some of the databases will also contain results from patient tests with gloves but not separated from other test results. The different data systems used for storage of dermatological test data and communication facilities are thoroughly presented in the publication Seminars on Dermatology, June, 1989, Computers and Dermatology.[57]

REFERENCES

1. Roder, M. M., A guide for evaluating the performance of chemical protective clothing (CPC). *DHHS (NIOSH) Publication No. 90-109,* Publications Dissemination, DSDTT, NIOSH, U.S.A., 1990.

2. Laursen, H., Prövning af beskyttelsehandskers resistense mod organiske opplösningsmiddler. II. *Arbejdmiljöfondets forskningsrapporter,* Arbejdsmiljöfondet, Copenhagen, 1985 (in Danish).
3. Henriksen, H. R. and Petersen, H. J. S., Protective clothing against chemicals. Better gloves against epoxy products and other chemicals, *Arbejdsmiljöfondets forskningsrapporter,* Arbejdsmiljöfondet, Copenhagen, 1987 (in Danish).
4. Nelson, G. O., Lum, B. Y., Carlson, G. J., Wong, C. M., and Johnson, J. S., Glove permeation by organic solvents, Lawrence Livermore Laboratory, UCRL-81766 Rev. 1, Preprint, 1979.
5. Nolen, R. L., Glove permeation/absorption, Appendix X, National Toxicology Program NTP, Radian Corporation, Austin, TX, 1984.
6. Conoley, M., Prokopetz, A. T., and Walters, D. B., Permeation of chemical protective clothing. II. Cumulative permeation test results for the National Toxicology Program, U.S. Department of Commerce, National Technical Information Service, NTIS, Radian Corporation, Austin, TX, 1990.
7. Mellström, G. A., Protective gloves of polymeric materials: experimental permeation testing and clinical study of side effects, *Arbete och Hälsa 1991:10,* Arbetsmiljöinstitutet, Stockholm, 10, 1991 (in English).
8. Barker R. L. and Coletta, G. C., Eds., *Performance of Protective Clothing: First Symposium,* ASTM STP 900, American Society of Materials and Testing (ASTM), Philadelphia, 1984.
9. Mansdorf, S. Z., Sager, R., and Nielsen A., Eds., *Performance of Protective Clothing: Second Symposium,* ASTM STP 989, American Society of Materials and Testing (ASTM), Philadelphia, 1988.
10. Perkins J. L. and Stull, J. O., Eds., *Chemical Protective Clothing in Chemical Emergency Response,* ASTM STP 1037, American Society of Materials and Testing (ASTM), Philadelphia, 1989.
11. McBriarty, J. P. and Henry, N. W., Eds., *Performance of Protective Clothing: Fourth Volume,* ASTM STP 1133, American Society of Materials and Testing (ASTM), Philadelphia, 1992.
12. Mellström, G. A. and Carlsson B., Eds., *2nd Scand. Symp. Protective Clothing against Chemicals and Other Health Risks,* Solna, Stockholm, Sweden, Nov. 1986, *Arbete och Hälsa 1987:12, National Institute of Occupational Health,* Solna, NOKOBETEF, c/o HAZPREVENT, DK-1420 Copenhagen, Denmark, NOKOBETEF II, 1987 (in English).
13. Eggestad, J., Ed., *3rd Scand. Symp. Performance of Protective Clothing against Chemicals and Other Health Risks,* Gausdal, Norway, Sept. 1989, NOKOBETEF, c/o HAZPREVENT, DK-1420 Copenhagen, Denmark, NOKOBETEF III, 1989 (in English).
14. Eggestad, J., Ed., *3rd Scand. Symp. Performance of Protective Clothing against Chemicals and Other Health Risks,* Supplement volume, NOKOBETEF, c/o HAZPREVENT, DK-1420 Copenhagen, Denmark, NOKOBETEF III, 1989 (in English).
15. Mäkinen H., Ed., *4th Scand. Symp. Protective Clothing Against Chemicals and Other Health Risks,* Kittilä, Finland, Feb. 1992, NOKOBETEF, c/o HAZPREVENT, DK-1420 Copenhagen, Denmark, NOKOBETEF IV, 1992 (in English)
16. Edmont Chemical Resistance Guide, Form nr CRG-GC-883, Becton Dickinson and Company, U.S.A., 2nd ed., 1983.

17. Pioneer, Liquidproof industrial gloves, Pioneer Industrial Products, Willard, OH, 1986.
18. 4H Chemical protection list, Safety 4 A/S, Lyngby, Denmark, Nov. 1991.
19. High technology line. Setting standards for hand protection in clean rooms. Chemical resistance guide, High-Tech Products, Edmont Europe, Becton Dickinson, Belgium, Oct. 1986.
20. Siebe Norton permeation resistance guide, IH-244, Silver Shield gloves, Siebe North, Inc., North Hand Protection, Charleston, 1986.
21. Ansell/Edmont Industrial, Edmont Europa, Belgium, 1991.
22. Marygold Industrial Gloves, LRC Products LTD, London, 1991.
23. Goydan, R., Schwope, A. D., Loyd, S. H., Huhn, L. M., CPCbase, a chemical protective-clothing data base for the personal computer, *Performance of Protective Clothing: Second Symposium, ASTM STP 989,* Mansdorf, S. Z., Sager, R., and Nielsen, A. P., Eds., American Society for Testing and Materials, Philadelphia, 403, 1988.
24. Mellström, G. A., Lindahl, G., and Wahlberg, J. E., DAISY: Reference database on protective gloves, *Seminars in Dermatology: Computers and Dermatology,* Vol. 8, No. 28, 75, 1989.
25. Forsberg, K. and Keith, L. H., Chemical protective clothing performance index book, John Wiley & Sons, New York, 1989.
26. Schwope, A. D., Costas, P. P., Jackson, J. O., Stull, J. O., and Weitzman, D. J., *Guidelines for the Selection of Chemical Protective Clothing,* 3rd ed., Vol. I and II, American Conference of Governmental Industrial Hygienists, Cincinnati, 1987.
27. Forsberg, K. and Mansdorf, S. Z., Quick selection guide to chemical protective clothing, *Van Nostrand Reinhold,* New York, 1989.
28. Johnson, J. S. and Anderson, K. J., Eds., *Chemical Protective Clothing,* American Industrial Hygiene Association, Vol. I and II, Akron, OH, 1989.
29. Keith, L. C. and Walters, D. B., Eds., *The National Toxicology Program's Chemical Data Compendium,* Lewis Publishers, Chelsea, MI, 1992.
30. Kotilainen, H. R., Avato, J. L., and Gantz, N. M., Latex and vinyl nonsterile examination gloves: status report and biological methods, *Appl. Environ. Microbiol.,* June, 1627, 990.
31. Kotilainen, H. R., Brinker, J. P., Avato, J. L., and Gantz, N. M., Latex and vinyl examination gloves. Quality control procedures and implications for health care workers, *Arch. Intern. Med.,* 149, 2749, 1989.
32. Kotilainen, H. R., Cyr, W. H., Truscott W., Gantz, N. M., Routson, L. B., and Lytle, C. D., Ability of 1000 ml water leak test for medical gloves to detect gloves with potential for virus penetration, *Performance of Protective Clothing: Fourth Volume, ASTM STP 1133,* American Society of Materials and Testing (ASTM), Philadelphia, 38, 1992.
33. Klein, R. C., Party, E., and Gershey, E. L., Virus penetration of examination gloves, *Biotechniques,* 9, 196, 1990.
34. Dalgleish, A. G. and Malkovsky, M., Surgical gloves as a mechanical barrier against human immunodeficiency viruses, *Br. J. Surg.,* 75, 171, 1988.
35. Matta, H., Thompson, A. M., and Rainey, J. B., Does wearing two pairs of gloves protect operating theatre staff from skin contamination?, *BMJ,* 297, 597, 1988.

36. Gleich, P., Probleme mit Einmalhandschuen (Problems with examination gloves), *Hygiene + Medizin, AUS DER PRAXIS,* 448, 1987.
37. Yangco, B. G., Yangco, N. F., What is leaky can be risky: a study of the integrity of hospital gloves, *Infect. Control Hosp. Epidemiol.,* 10, (12), 553, 1989.
38. Zbitnew, A., Greer, K., Heise-Qualtiere, J., and Conly, J., Vinyl versus latex gloves as barrier to transmission of viruses in the health care setting, *J. Acquired Immune Deficiency Syndromes,* 2, 201, 1989.
39. Gonzales, E. and Naleway, C., Assessment of the effectiveness of glove use as a barrier technique in the dental operatory, *JADA,* 117, 467, 1988.
40. Reingold, A. L., Mark, A. K., and Hightower, A. W., Failure of gloves and other protective devices to prevent transmission of hepatitis B virus to oral surgeons, *JAMA,* 259, No. 17, 2558, 1988.
41. Carey, R., Herman, W., Herman, B., Krop, B., and Casamento, J., A laboratory evaluation of standard leakage tests for surgical and examination gloves, *J. Clin. Engin.,* 14, 133, 1989.
42. Korniewicz, D. M., Laughon, B. E., Cyr, W. H., Lytle, C. D., and Larson, E., Leakage of virus through used vinyl and latex examination gloves, *J. Clin. Microbiol.,* April, 787, 1990.
43. Paulssen, J., Eidem, T., and Kristiansen R., Perforation in surgeons' gloves, *J. Hosp. Inf.,* 11, 82, 1988.
44. Korniewicz, D. M., Laughon, B. E., Butz, A., and Larson, E., Integrity of vinyl and latex gloves, *Nurs. Res.,* 38, 144, 1989.
45. Marin, J., Dragas, A. Z., and Mavsar, B., Virus permeability of protective gloves used in medical practice, *Zbl. Hyg.,* Gustav Fischer Verlag/Stuttgart, New York, 191, 516, 1991.
46. Glove update (FDA report), Fred Sadler, The Freedom of Information Office HFZ-82, Rockville, MD, 1989.
47. SPRIMA-TEST 1987 and 1991. Examination gloves for single use, SEMKO AB, Medical Devices, (Kista) Stockholm, Sweden, 1991 (in Swedish).
48. SPRIMA-TEST 1982/83 and 1990. Sterile surgeon gloves, SEMKO AB, Medical Devices, (Kista), Stockholm Sweden, 1990 (in Swedish).
49. Cronin, E., *Contact Dermatitis,* Churchill Livingstone, Edinburg, 1980.
50. Adams, R. M., *Occupational Skin Disease,* Grune & Stratton, New York, 1983.
51. Maibach, H. I., Ed., *Occupational and Industrial Dermatology,* 2nd ed., Year Book Medical Publishers, Chicago, 1986.
52. Rycroft, R. J. G., Menné, T., Frosch, P. J., and Benezra, C., Eds., *Text Book of Contact Dermatitis,* Springer-Verlag, Berlin, 1992.
53. Nutt, A. R., *Toxic Hazards of Rubber Chemicals,* Elsevier Applied Science Publishers, London, 1984.
54. Turjanmaa, K., Latex glove contact urticaria, *Acta Universitatis Tamperensis ser A,* (Academic dissertation), University of Tampere, Finland, Vol. 254, 1988.
55. Estlander, T., Occupational skin diseases in Finland, 1974-1988, (Academic dissertation), *Acta Derm-Vener,* Suppl. 155, 1990.
56. Sensitivity to latex in medical devices, in *Proc. Int. Latex Conf.,* Baltimore, November 5-7, Food and Drug Administration, Centers for Disease Control, National Institute of Allergy and Infectious Diseases, Nov. 1992.
57. Dooms-Goossens, A., Ed., Computers in Dermatology, in *Seminars in Dermatology,* W. B. Saunders Company, Philadelphia, 8, No. 2, 1989.

INDEX

INDEX

A

Accelerators, 27, 28; see also Allergens
Acceptable quality level (AQL), 72, 111–112
Acquired immune deficiency syndrome (AIDS), 48, 110; see also Human immunodeficiency virus (HIV)
Airborne chemical exposures, 13, 14
Air leak tests, 72, 87–88, 112
Allergens; see also Occupational dermatitis, Type I hypersensitivity, Type IV hypersensitivity
   chemical additives, 133–134, 175–176, 197–198, 225–228
   glove powder, 145–146, 172, 193, 199, 249
   latex proteins, 134–135, 198, 199–200, 249–250
Allergic contact dermatitis; see Contact dermatitis
Allergic contact eczema; see Contact dermatitis
Allergies, delayed–type; see Type IV hypersensitivity
Allergies, immediate–type; see Type I hypersensitivity
Allergies, latex glove; see Latex allergies
Allergy testing
   glove use, 142, 245
   intradermal, 142–143
   patch, 138–139, 207–212
   radioallergosorbent (RAST), 143–145, 245
   risks, 138, 141–142, 188, 245
   skin prick, 139–142, 243–244
   with used gloves, 165, 217–218
American National Standards Institute (ANSI), 46
American Society for Testing and Materials (ASTM), 40, 46; see also ASTM standards, ASTM test cells
AMK test cells, 66
Anaphylactic reactions, 135, 136, 199, 201
   allergy testing and, 138, 142, 188, 245
   intraoperative, 136, 194–195
Animal studies, 3–4, 92–106
ANSI; see American National Standards Institute
Antineoplastic drugs, 145
Antioxidants, 27; see also Allergens
Antiseptics, 145

AQL; see Acceptable quality level
ASTM; see American Society for Testing and Materials
ASTM standards
   ASTM D 3577, rubber surgical gloves, 48
   ASTM D 3578, rubber examination gloves, 48
   ASTM D 5151, hole detection in medical gloves, 48
   ASTM ES 21, resistance to synthetic blood, 48
   ASTM ES 22, resistance to viral penetration, 48
   ASTM F 739–91, permeation test method, 47, 63, 64
   ASTM F 903–87, penetration test method, 71, 72
   ASTM F 1001–89, test chemical selection, 47
   ASTM F 1383, intermittent contact, 70
ASTM test cells, 54–55, 58–59, 65–66
Atopy, latex allergy and, 136, 171, 195, 216, 248

B

Bacterial endotoxins, 172, 261–262
Bacteriophage
   properties, 118
   virus penetration tests using, 119, 121
BDT; see Breakthrough detection time
Benzoyl peroxide, 228
Black–rubber mix, 191–193, 197
Blended polymer gloves, 31
Breakthrough detection time (BDT), 56
Breakthrough time (BT)
   defined, 56
   factors affecting, 60–63, 66
   glove selection and, 105, 272
   testing, 47, 55, 66, 86
n–Butanol, 94–95, 98–105
Butylhydroxyanisole (BHA), 228
Butyl rubber gloves
   permeation tests, 92–93, 95–98
   uses, 23
   in vivo tests, 94–95, 98–105

C

Carbamates, 134
Carba mix, 139, 166, 175, 197

CBS; see N-Cyclohexylbenzothiazyl sulfenamide
CE mark, 41
CEN; see Committee for European Normalization
Cetyl pyridinium chloride, 228
Chemical exposure assessment, 15
Chemical fill tests, 112
Chemical leukoderma; see Leukoderma, chemical
Chemical permeation testing; see Permeation testing
Chemical–protective suits, field testing, 87–88
Chemical resistance testing; see Degradation testing, Permeation testing, Penetration testing
Chem Plus Glove, 22
Chlorination, 30–31
Chloroprene rubber gloves, 24
Cholinergic urticaria, 259
Chromium, 228
Closed–loop test systems
 breakthrough time and, 66
 detection limits, 63
Committee for European Normalization (CEN), 39–40
Composite polymer gloves, 31
Contact dermatitis; see also Glove dermatitis, Irritant contact dermatitis, Occupational dermatitis, Type IV hypersensitivity
 clinical diagnosis of, 135–147, 161–166, 243–245
 contact urticaria and, 174
 immediate, 189, 196, 198
 occupational causes of, 145, 217–218, 223–224
 prevention of, 7–8
 types of, 132–135
Contact urticaria; see also Latex allergies, Type I hypersensitivity
 allergologic evaluation of, 187–189
 associated symptoms, 136–137, 172–174, 242–243, 245–247
 contact dermatitis and, 174
 defined, 242
 nonimmunologic, 258–259
 prevalence of, 247–248
 prevention of, 250–251
 risk factors for, 248
Contact urticaria syndrome, 242–243
Cornstarch allergies, 145–146, 172, 249

Corrosion, skin, 14
Cut–resistant gloves, 32–33, 290
N–Cyclohexylbenzothiazyl sulfenamide (CBS), 226

D

Damaged skin, percutaneous absorption and, 94, 98–100, 104–105
Decontamination, 18–19, 80, 278–279
Defatting, skin, 91
Defects (holes)
 detection, 114–116
 leakage rate, 115–116
 viral infection and, 121–122, 289–290
Degradation, chemical permeation and, 62, 101–103, 105
Degradation testing, 54–55, 80–82
 data sources, 294–297
Delayed allergic reactions; see Type IV hypersensitivity
Depigmentation, skin, 259–261
Dermatitis; see Contact dermatitis, Glove dermatitis, Irritant contact dermatitis, Occupational dermatitis, Type IV hypersensitivity
Diamine mix, 139
Dibenzothiazyl disulfide (MBTS), 226
Diethyldithiocarbamate zinc (ZDC), 26, 191, 195, 197
Diffusion, Fick's first law of, 60, 104
Dipentamethylenethiuram disulfide, 226
Diphenylguanadine, 166
Diphenylthiourea, 26
Disposable gloves, 21, 22, 285–286
Documentation of Biological Exposure Indices, 14
Domestic gloves, 22
Doses, chemical exposure, 13–15
Double–gloving; see also Glove liners
 skin irritation and, 256
 surgery and, 120, 289–290

E

Eczema, hand, 136, 171, 216, 234
Edema, solvents and, 91
Elastyren® gloves, 22, 24
Endotoxins, bacterial, 172, 261–262
Ethylene chlorhydrin, 263
Ethylene glycol, 263
Ethylene-methylmethacrylate gloves, 24

Index

Ethylene oxide (ETO), 133, 172, 249, 262–263
European standards
 penetration testing, 71–72
 permeation testing, 57, 58, 59, 60, 63
 protection index, 70
 protective gloves, 39–43
Examination gloves; see also Medical gloves
 failure rate of, 120–121
 virus penetration tests of, 121
Exposure, chemical
 airborne, 13, 14
 assessment analysis, 15–16, 216–217, 271–272
 doses, 13–15
 recommendations, 14–15

F

Fabric gloves
 formaldehyde and, 211
 manufacturing, 32–33
Fatigue, latex, 123–124
FDA; see U.S. Food and Drug Administration
FDA 1000 ml water leak test
 protocol, 110–112
 sensitivity of, 116
 virus penetration and, 121–122
Fick's first law, 60, 104
Field test methods, 80–88
Fill tests
 air, 72, 87–88, 112
 chemical, 112
 water, 71, 72, 110–112
Flow rate, permeation testing and, 64–65
Fluorocarbon rubber gloves, 24
Foil gloves, 33
Formaldehyde, 145, 211
Formers, glove, 25, 31, 32
4H™–glove, 25
Fruit allergies, latex allergy and, 136, 174, 189, 196

G

Glove dermatitis; see also Contact dermatitis, Immediate contact dermatitis, Occupational dermatitis, Type IV hypersensitivity
 chemical penetration and, 176–177, 207–212

clinical management of, 147–149
 diagnosis of, 232–233
 exposure factors, 216–217
 glove powder and, 145–146, 172, 193, 199, 249
 prevention of, 233–234
 symptoms of, 136–137, 216, 230
Glove liners, 32–33, 149, 211, 233, 290
Glove powder
 airborne contact with, 172, 243
 allergies, 145–146, 193, 197, 199, 249
 glove dermatitis and, 132, 145–146, 216, 249
Gloves; see Protective gloves, and under glove material (e.g., Latex gloves)
Grip, 31
Guide for the Selection of Chemicals to Evaluate Clothing Materials, 47
Guide to Classification and Marking of Hazardous Substances, 14
Guinea pigs, *in vivo* tests on, 92, 94–95, 98–106

H

HA; see High-ammonia latex
Hand eczema, latex allergy and, 136, 171, 216, 234
Hazards, evaluation of, 12–17, 271–272
HBV; see Hepatitis B virus (HBV)
Health care workers, 135, 168, 172, 247, 285–287; see also Medical gloves, Occupational dermatitis
Hepatitis B virus (HBV)
 glove punctures and, 122
 properties of, 118
 protective gloves and, 110, 289–290
Herpes simplex virus (HSV)
 properties of, 118
 protective gloves and, 110
 virus penetration tests, 121
Hevea brasiliensis, 25, 134
Hevein, 134
High–ammonia (HA) latex, 25, 26
HIV; see Human immunodeficiency virus (HIV)
Household gloves, 21, 22
HSV; see Herpes simplex virus
Human immunodeficiency virus (HIV)
 properties, 118
 protective gloves and, 48, 110, 289, 290
 punctured gloves and, 122

virus penetration tests, 119, 121, 290
Hypersensitivity; see Type I hypersensitivity, Type IV hypersensitivity
Hypoallergenic gloves
   labeling, 34, 149
   types, 22, 24, 148–152

I

IgE; see Immunoglobulin E
Immediate allergic reactions; see Type I hypersensitivity
Immediate contact dermatitis, protein–induced, 189, 196, 198
Immersion test, 54
Immunoglobulin E (IgE)
   radioallergosorbent testing (RAST), 143–145, 245
   type I hypersensitivity and, 134, 174, 195
   type IV hypersensitivity and, 195
Industrial gloves
   defined, 22
   industrial hygiene assessment and, 11–19
   selection, 270–275, 298
Industrial hygiene
   process, 12
   risk assessment, 11–19
Infrared gas analyzer, 84–86
Inspection, glove, 18–19, 80
Integrity testing, medical glove, 110–122
International Standards Organization (ISO), 40
Intradermal testing, 142–143
Irritant contact dermatitis; see also Contact dermatitis
   allergen sensitization and, 146, 165, 171, 230
   atopic patients and, 255
   causes, 91, 132–133, 255–258
   clinical diagnosis, 217–219, 257–258
   patch testing and, 218
ISO/DIS test cell, 59
N–Isopropyl-N'-phenyl-p-phenylenediamine (IPPD), 191
ISO; see International Standards Organization
ISO standards
   permeation testing, 57, 58, 59, 63
   protective gloves, 40

K

Kevlar® glove liners, 32–33
Knitted gloves, 32–33

L

LA; see Low-ammonia latex
Labeling, glove, 34, 148, 149, 251
Latex allergies; see also Allergens, Contact dermatitis, Contact urticaria, Type I hypersensitivity, Type IV hypersensitivity
   fruit allergies and, 136, 174, 189, 196
   inhalant, 193–194
   management, 147–149
   occupation and, 167–168, 229, 247
   prevalence, 135–136, 162–163, 196, 200–202, 223–224
   reactions, 132, 133–135, 245–246
   risk factors, 135–136, 200–202, 229–230, 248
Latex gloves; see also Allergens, Medical gloves
   manufacturing, 25–32
   material fatigue, 123–124
   permeation tests, 92–93, 95–98
   pores, 123–124
   storage conditions, 124–125
   uses, 23
   virus penetration tests, 119–121
   in vivo tests, 94–95, 98–105
Latex proteins
   allergic reactions and quantity of, 159–161
   extraction techniques, 142, 188, 244
   reduction, 152, 158
Leaching, in glove manufacturing, 30
Leakage tests
   air, 72, 87–88, 112
   water, 71, 72, 110–112
Leather gloves, 32, 172
Leukoderma, chemical
   animal studies and, 261
   chemicals causing, 260–261
   differential diagnosis, 261
   symptoms, 259–260
LifeLiner™ gloves, 32
Low–ammonia (LA) latex, 25
Lycra® glove liners, 32

M

Maceration, gloves and, 132, 256
Manufacturing, protective glove, 25–34
MBMBP; see 2,2'-Methylene-*bis*(4-methyl-6-*tert*-butylphenol)
MBTS; see Dibenzothiazyl disulfide
MDPR; see Minimum detectable permeation rate

Index

Medak™ gloves, 32
Medarmor™ Cut–and–Puncture–resistant gloves, 32–33
Medical gloves
  categories, 286
  cut–resistant, 32–33, 290
  double–gloving, 289–290
  examination gloves, 120–121
  penetration testing, 72
  perforation, 120, 289, 290
  quality control, 72, 110–112, 125
  selection, 284–288
  standards, 48
  storage, 288
  test methods, 112–116
  transmission of infection and, 288
  used, 119–121
  virus penetration, 289
Mercaptobenzothiazole, 139, 166
Mercapto mix, 139, 166
  allergy testing and, 197
2,2'–Methylene–bis(4–methyl–6–tert–butylphenol) (MBMBP), 227
Methyl methacrylate, 145
Minimum detectable permeation rate (MDPR), 57, 63
MMBT; see Morpholinyl mercaptobenzothiazole
Modified Human Draiz Test, 34, 149
Molds, glove, 25, 31, 32
Morpholinyl mercaptobenzothiazole (MMBT), 191, 226

N

National Institute for Occupational Safety and Health, 12
Natural latex gloves; see Latex gloves
Natural rubber gloves; see Latex gloves
Needle–sticks, medical gloves and, 120, 289, 290; see also Glove liners
Neoprene® gloves, 24
Nitrile rubber gloves, 24, 122
Normalized breakthrough detection time (NBDT), 56, 63
North American Contact Dermatitis Group, 139

O

Occ–Derm project, 167–168
Occlusion
  effects on skin, 92, 101, 104
  glove use and, 256
Occupational dermatitis, 145–146, 167–168, 171–179, 195
  glove contamination and, 174, 211
Occupational Safety and Health Administration (OSHA)
  Permissible Exposure Limits (PELs), 14
Open–loop test systems
  breakthrough time and, 66
  detection limits, 63
OSHA; see Occupational Safety and Health Administration
Ozone, latex and, 124–125

P

Patch testing
  false negatives, 138
  gloves plus allergens, 208–212
  indications, 136, 138
  risks, 138
  standard chemicals, 138–139
  technique, 138–139
  with used gloves, 165, 217–218
PE; see Polyethylene
PELs; see Permissible Exposure Limits
Penetration testing, 71–72, 82, 117
  data sources, 298–299
Percutaneous absorption
  animal studies, 92–106
  defined, 3
  factors affecting rate, 101
  measuring rate of, 94
Perforation, surgical glove, 120, 289; see also Defects (holes)
Permeation, chemical, 3
Permeation rate (PR)
  behavior, 62, 101–103
  defined, 56
  glove materials and, 67–68
  latex glove flexing and, 123
  temperature and, 67
Permeation testing
  ASTM F 739–91, 47, 55
  data sources, 294–297
  detector tube method, 86
  factors affecting, 60–69
  gas or vapor, 84–86
  highly toxic substances, 70
  intermittent contact, 69–70
  ISO/DIS, 66–67

low volatility chemicals, 70–71
low water solubility chemicals, 70–71
non–volatile liquids, 84
permeation cup, 86–87
sampling methods, 63, 66
volatile liquids, 83–84
Permissible Exposure Limits (PELs), 14
Perry® Cut–Resistant gloves, 32
Photoionization detector, 84, 87
Physical hazards assessment, 16
Piperidine, 227
Piperidine pentamethylenedithiocarbamate, 227
Plastic gloves; see also Polyvinylchloride
 (PVC) gloves, Polyethylene (PE) gloves
 allergies, 222, 223
 manufacturing, 32, 33
 types, 24–25
Poiseuille's equation, 115
Polyethylene (PE) gloves
 uses, 24
 virus penetration tests, 119
cis–1,4 Polyisoprene, 134
Polyvinylalcohol gloves, 25
Polyvinylchloride (PVC) gloves
 allergies, 199, 224, 228–229
 effects of solvents on, 103
 leakage tests, 121
 permeation tests, 92–93, 95–98
 punctures, 122
 uses, 25
 virus penetration tests, 119, 121, 122
 in vivo tests, 94–95, 98–105
PR; see Permeation rate
Prenyltransferase, 134
Pressure testing
 chemical–protective suits, 87–88
 materials, 82
Pressure urticaria, 258–259
Protection index, 70
Protective gloves; see also Glove liners,
 Protective glove testing, Side effects, and
 specific types of gloves
 comfort, 277
 decontamination, 18–19, 80, 278–279
 dexterity, 277–278
 disposable, 21, 22, 285–286
 general rules, 279–280
 hazard assessments and, 271–272
 health care workers and, 285–287
 household (domestic), 21, 22
 hypoallergenic, 22, 24, 148–152
 industrial, 22
 industrial hygiene assessments, 11–19

inspection, 18–19, 80
labeling, 34, 148, 149, 251
lined, 148
manufacturing, 25–34
materials used, 22–25
quality vs. cost, 284–285
reuse, 80, 274–275
risks, 17–18
selection, 11–19, 270–275, 284–290, 298
special, 22
standards, 39–43, 46–49
storage conditions, 27, 33
suitability assessment, 276–278
thickness, 22, 93, 95, 103–104, 105
training and, 18–19, 279
types, 21–22
in vivo testing, 92, 94–95, 98–106
Protective glove selection
 industrial, 270–275
 industrial hygiene assessment and, 11–19
 medical, 284–288
 published guidelines, 298
Protective glove testing
 degradation, 54–55, 80–82
 penetration, 71–72, 82, 114–122
 permeation, 55–69, 82–87
 sensitivity, 114–116
 standard performance tests, 46–49
 in vivo, 92, 94–95, 98–106
 whole–glove, 87
Protective glove use
 comfort and, 277
 dexterity and, 277–278
 general rules, 279–280
 health care workers, 285–287
 inspection and, 80
 training and, 279
Provocative allergy testing
 false positives, 165
 with gloves, 165–166
 precautions, 158–159, 188–189
 risks, 138
Puncture–resistant gloves, 32–33
Punctures, viral infection and, 121–122; see
 also Defects (holes)
PVC gloves; see Polyvinylchloride (PVC)
 gloves

Q

*Quick Selection Guide to Chemical
 Protective Clothing*, 14

Index

R

Radioallergosorbent test (RAST), 143–145, 245
Relative absorption rate, 94–95
Relative permeation rate, 93
Repel™ glove liners, 32
Retroviruses; see Human immunodeficiency virus (HIV)
Risk assessment
  hazard evaluation, 12–14, 271–272
  work practices and, 17
Risk/benefit evaluation, 17–18
Rubber chemicals, 28, 133–134, 197–198, 225–228
Rubber elongation factor, 134
Rubber glove allergies; see Latex allergies
Rubber gloves, 23–24; see also Butyl rubber gloves, Latex gloves, Nitrile rubber gloves
Rubber latex gloves; see Latex gloves
Rubber tree, 25

S

Sempermed® Protector, 22
Sewn gloves, 32–33
Side effects, 8, 23, 33–34; see also Glove dermatitis, Latex allergies, Skin
  published reports, 299–300
Silver Shield™, 25
Skin
  corrosion, 14
  defatting, 91
  depigmentation, 259–261
  irritation, 94, 98-100, 132–133, 216–219, 255–258
  solvents and, 91
Skin notations, 14
Skin prick testing
  anaphylactic reactions, 142
  risks, 141, 244
  technique, 139–142, 244
Solvents
  effects on skin, 91
  glove manufacturing and, 31
  percutaneous absorption, 92–106
  systemic effects, 91
Sorbic acid, 189, 228, 233
Special gloves, 22
Spectra® Fiber gloves, 32
Spina bifida, latex allergy and, 136, 195, 201
Splash testing, 69–70

SPR; see Steady-state permeation rate
Standards; see European standards, ISO standards, U.S. standards
Standard test methods
  medical glove integrity, 110–116
  protective clothing, 57–58
Steady–state permeation rate (SPR), 93, 101
Stokes radii, 118
Storage conditions, 27, 33
Styrene-butadiene rubber gloves, 24
Styrene-ethylene-butadiene rubber gloves, 24
Sulfur, 26–27, 28
Supported gloves, 32
Surface drag, reduction of, 30, 31
Surface tension effects
  hole detection and, 114–116
  microsphere penetration tests and, 117
Surgical gloves; see Medical gloves

T

Tactylon™ gloves, 22, 24
Test cells
  alternative, 59, 66
  AMK, 66
  ASTM, 58–59, 65–66
  ISO/DIS, 59
Test data, published, 293–300
Tetraethylthiuram disulfide, 226
Tetramethylthiuram disulfide (TMTD), 26, 191, 226
Tetramethylthiuram monosulfide (TMTM), 191, 226
Thiazoles, 134
4,4'–Thiobis(6–*tert*–butyl–*meta*–cresol), 227, 228
Thioureas, 26, 134
Thiuram–free gloves, 34
Thiuram mix, 139, 166, 175, 197–198
Thiurams, 134, 175
Threshold Limit Values for Chemical Substances and Physical Agents, 14
TMTD; see Tetramethylthiuram disulfide
TMTM; see Tetramethylthiuram monosulfide
Toluene, 94–95, 98–105
Training, 18–19, 279
Transappendageal transport, 3
Transepidermal transport, 3
Transmembrane pressure
  flow through leaks and, 117
  surgical gloves and, 117
1,1,1–trichloroethane, 94–95, 98–105
Type I hypersensitivity, see also Contact urticaria

chlorination and, 31
clinical manifestations, 136, 193–194
diagnostic methods, 243–245
differential diagnosis, 145–147, 243–245
latex protein and, 134–135, 198–200
leaching and, 30
occupational risk factors, 135–136
skin prick testing and, 139, 243–244
use testing and, 136–137
Type IV hypersensitivity; see also Contact dermatitis
differential diagnosis, 145–147
patch testing and, 138
risk factors, 135, 190, 229–230
skin prick testing, 139
symptoms, 133

U

Universal precautions, medical, 284
Urticaria; see Cholinergic urticaria, Contact urticaria, Pressure urticaria
U.S. Food and Drug Administration (FDA), 34, 111–112
U.S. standards; see also ASTM standards
intermittent chemical contact, 70
medical gloves, 48, 49
penetration testing, 71, 72
permeation testing, 47, 55, 57–59, 60
Use testing, glove allergies and, 142, 245

V

Vinyl gloves; see Polyvinylchloride (PVC) gloves
Viruses
glove penetration, 72, 110, 116–122, 289, 290
physical properties, 118
Viton® gloves, 24
Vulcanization, 26–27

W

Water leak tests, 71, 72, 110–112
Welded–seam gloves, 33
Worker training
glove use and, 279
risk reduction and, 18–19
Work practices, evaluation of, 17, 270
Wounds, percutaneous absorption and, 94, 98–100, 104–105

Z

Zinc dibutyldithiocarbamate (ZDBC), 26, 227
Zinc diethyldithiocarbamate (ZDC), 26, 227
Zinc dimethyldithiocarbamate, 227
Zinc 2–mercaptobenzothiazole (ZMBT), 26
Zinc oxide, 27, 28